T0140340

Progress in IS

Preface

Sustainable development is the idea that human societies must survive and meet their needs without compromising the ability of future generations' survival. Specifically, sustainable development is a way of organising society so that it can exist in the long term. This means taking into account both the imperatives of the present and those of the future, such as the preservation of the environment and natural resources or social and economic equity. The aim of this book, **Sustainable Development through Data Analytics and Innovation—Techniques, Processes, Models, Tools, and Practices**, is to contribute to the research and knowledge on sustainable development and the role of business, innovation, and digital technologies in achieving the Sustainable Development Goals (SDGs).

The chapters highlight areas of interest that draw the link between sustainable development and digital technologies. The book addresses topics such as achieving sustainable development through green technologies, thereby highlighting how the concepts and practices of green technologies could create sustainable opportunities for business and, likewise, organisations. In contemporary times, digital technologies are driving by digital functionalities such as Blockchain and its improving innovative use across multiple sectors beyond finance and banking. To achieve more efficient, effective, and sustainable business decisions, organisations heavily rely on the use of data to improvise their current situation and environment as well as predict likely outcomes based on decisions taken today. Hence, data science and data analysis have become increasingly applied in sustainable development. These technologies enable organisations derive better insights into the data as well as the ability to prescribe, diagnose, and predict the sustainable impact of our current business activities.

While digital technologies support business and organisational operations to achieve business and sustainable objectives, there is a human side to technologies that must be considered. Hence, the book includes chapters that draw a relationship between the users of technologies and the technological tools. An instance is the relationship between digitalisation and passenger air travel, as well as the effect of green purchasing behaviour on social media. Also, reimaging corporate reporting in the digital age and a look into the responses to the global Covid-19 pandemic.

Finally, the book includes chapters on the relationships between business analytics, education, innovation industrial development, and sustainability. In line with the objectives, the book highlights key issues in our business environment and to instigate further research and knowledge in sustainability and its applicability to all aspects of life.

Oldenburg, Germany   Jorge Marx Gómez
Sharjah, United Arab Emirates   Lawal O. Yesufu
June 2022

Contents

Contributors

Shahira O. Abdalla Department of Human Resource Management, Faculty of Business, Higher Colleges of Technology, Abu Dhabi, UAE

Sami Alajlani Department of Finance, Faculty of Business, Higher Colleges of Technology, Abu Dhabi, UAE

Shahira El Alfy Faculty of Business, Higher Colleges of Technology, Abu Dhabi, UAE

Safwat Al Tal Department of Business Analytics, Faculty of Business, Higher Colleges of Technology, Abu Dhabi, UAE

Wajeeha Aslam Department of Business Administration, IQRA University, Karachi, Pakistan

Cees Bil RMIT University, Melbourne, VIC, Australia

Benjamin Silas Bvepfepfe Faculty of Business, Higher Colleges of Technology, Abu Dhabi, UAE

Kashif Farhat Department of Marketing, Mohammad Ali Jinnah University, Karachi, Pakistan

Ahmad Faisal Hayek Department of Accounting, Faculty of Business, Higher Collogues of Technologies, Abu Dhabi, United Arab Emirates

Iryna Heiets RMIT University, Melbourne, VIC, Australia

Hanafi Al Hijazi Faculty of Computing and Informatics, University of Malaysia of Sabah, Kota Kinabalu, Sabah, Malaysia

Jiezhuoma La RMIT University, Melbourne, VIC, Australia

Kennedy Modugu Faculty of Business, Higher Colleges of Technology, Abu Dhabi, UAE

Puteri N. E. Nohuddin Faculty of Business, Sharjah Women's Campus, Higher Colleges of Technology, Abu Dhabi, UAE
Institute of IR4.0, National University of Malaysia, Bangi, Malaysia

Nora Azima Noordin Faculty of Business, Higher Colleges of Technology, Sharjah Women's Campus, Sharjah, UAE
Faculty of Business, Sharjah Women's Campus, Higher Colleges of Technology, Abu Dhabi, UAE

Shatha M. Obeidat Department of Management and Marketing, Faculty of Business, University of Qatar, Doha, Qatar

Marja Azlima Omar Faculty of Social Sciences and Humanities, University of Malaysia of Sabah, Kota Kinabalu, Sabah, Malaysia

Lucio Poma Department of Economics and Management at the University of Ferrara, Ferrara, Italy

Almaz Sandybayev Faculty of Business, Higher Colleges of Technology, Abu Dhabi, UAE

Haya Al Shawwa Business Division at the Higher Colleges of Technology, Sharjah, United Arab Emirates

Ilaria Vesentini Manufacturing Economic Studies (MECS), Modena, Italy

Lawal O. Yesufu Department of Business Analytics, Faculty of Business, Higher Colleges of Technology, Abu Dhabi, UAE

Zuraini Zainol Department of Computer Science, National Defence University Malaysia, Kuala Lumpur, Malaysia

The Impact of Business Analytics on Industry, Education, and Professional Development

Lawal O. Yesufu and Sami Alajlani

1 Introduction

Private and public companies hold huge amounts of data about their suppliers, services, products, and customers. According to Bayrak (2015), most businesses apply Business Analytics to utilize the technical data they store in data warehouses and databases by transforming it into useful operational insights. Today, Business Analytics is becoming an integral part of many investments, particularly because of its importance in decision-making activities (Brock & Khan, 2017). Furthermore, Business Analytics (BA) has an empowerment potential for investors due to its support when making tactical, operational, and strategic conclusions. Given the competitiveness in the market, some organizations are established entirely on Business Analytics concepts and depend on the ability to gather, analyze, and leverage data to influence success. The world is becoming greatly interconnected, whereby people use computers to collect volumes of data and enhance its accessibility. It is worth noting that data is reshaping human beings' perspectives about the world in recent decades (Bayrak, 2015).

New and sophisticated terms such as artificial intelligence (AI), the Internet of Things (IoT), and machine learning (ML) are becoming common where there is big data. Yin and Fernandez (2020) affirm that Business Analytics is one of the big data setting where buzzwords such as AI are applied. Yin and Fernandez (2020) support

L. O. Yesufu (✉)
Department of Business Analytics, Faculty of Business, Higher Colleges of Technology, Abu Dhabi, UAE
e-mail: Lyesufu@hct.ac.ae

S. Alajlani
Department of Finance, Faculty of Business, Higher Colleges of Technology, Abu Dhabi, UAE
e-mail: salajlani@hct.ac.ae

© The Author(s), under exclusive license to Springer Nature Switzerland AG 2022
J. Marx Gómez, L. O. Yesufu (eds.), *Sustainable Development Through Data Analytics and Innovation*, Progress in IS,
https://doi.org/10.1007/978-3-031-12527-0_1

1

their argument using the popularity curve in Google Trends that indicates increasing business analytics attention in the last decade. The growing demand for business analytics is due to the potential to revolutionize companies' data handling processes, such as collecting and analyzing user data from external sources like websites, mobile networks, data sensors, and social networks. Despite the increasing significance and popularity of the BA concepts, one can note that literature in the field is not abundant.

There is a limited systematic review of business analytics and studies that summarizes the concept. Many investors, the established and emerging business, need to understand BA in detail to apply it successfully in their corporations. This study intends to give an insight into Business Analytics, its relationship with other fields, and its applicability in education, businesses, and industries. The systematic review achieves the objective through a systematic review of literature from credible sources such as Scopus databases and Scopus Journals. After this introduction, the submission outlines the methodology used and the analysis procedure. The next sections include defining business analytics, its applications, business analytics relationship with data science and business intelligence (BI), and its uses in industrial and other investments. The study also evaluates the education and training in Business Analytics and a conclusion at the end.

2 Literature Review

The term business analytics has several definitions according to the literature reviewed in this paper. Scholars like Bayrak (2015) define it in relation to functions, and others focus on the significance and relationship with other disciplines such as data science and artificial intelligence. This paper analyses these different definitions and evaluates its operation to understand the meaning. Most importantly, the paper assesses the relationship between BA and common terms used in data settings like business intelligence and data science. Business analytics (BA) refers to the activity of gathering useful information from the stored corporate data and use it to enhance the efficiency and values. According to Yin and Fernandez (2020), Business analytics's primary purposes are collecting sufficient business data and leverage it in making critical decisions (El Alfy et al., 2019). Their definition implies that organizations can transform numerous data that they hold about their stakeholders, such as consumers and suppliers, and products and services, to make a critical decision (Vanani & Jalali, 2018). A further explanation of the definition is that the decision-making process at any given company works efficiently when the management utilizes accurate and reliable information than depending on personal knowledge or opinions. According to Bayrak (2015), Business analytics refers to an extensive assortment of technologies, procedures, and technologies used to collect, preserve, retrieve, and evaluate data to help organizations make sound decisions. The author argues that academics and investors use the terms business analytics

(BA), big data (BD), and business intelligence (BI) interchangeably to mean the same thing.

Business intelligence and business analytics are some of the frequently or commonly used terms in today's business environment. The terms apply in different settings whereby the business community uses business analytics while the informational technology community uses business intelligence (Mashingaidze & Backhouse, 2017). On other occasions, people use the two terms interchangeably to mean the same thing. Nevertheless, it is important to understand the similarities and differences between business intelligence and business analytics, because people have varying opinions. According to a study by Chae and Olson (2013), business analytics and business intelligence mean almost the same thing because they all describe analytics' potential to influence an effective decision-making process. The researcher implies that an individual can use any of the terms to illustrate the impacts of data analytics to business success (Priya et al., 2017). According to Sivarajah et al. (2017), people often use business analytics and data science to mean the same thing, but they are different domains (El Alfy et al., 2019). Unlike layman, professionals have a higher interest in differentiating the terms and use them where necessary because of the consequences in investments' performances. Data science refers to the study of data by applying statistics to offer important insights to investors without changing decisions (Sivarajah et al., 2017). On the other hand, business analytics is the evaluation of data to make fundamental decisions for a business. The definitions indicate that when one supports decision-making while the other gives insights to decision makers. In other words, the main difference between data science and business analytics is the scope of issues or corporate challenges that each addresses. Furthermore, data science integrates technology, statistical evaluation, and algorithms (Yesufu, 2021).

The concept aims to offer actionable intuitions of structured and unstructured information to address broader perspective concerns like unpredictable consumer behavior (Sivarajah et al., 2017). In contrast, business analytics is mainly about studying or evaluating structured business data to influence decision-making. Business analytics also solves a wide range of business problems and limiting factors. Using the terms interchangeably can influence undesirable outcomes because of their functional differences (Shen & Tzeng, 2016). The techniques used in business analytics are different from those of data science, and using them carelessly can yield undesirable and imperfect outcomes. Relying on the results derived using the wrong model or techniques can affect decision-making activity and corporate performance, competitiveness, and profitability. Data science is a broad field that covers all the activities designed to mine large databases. Wixom et al. (2013) argue that business intelligence stems from business analytics. Their main standpoint is that business analytics is a process, while BI is a source of insights. Gorman and Klimberg (2014) have a contrary opinion that views business analytics as an improvement or extension of business intelligence through operational research techniques and modern statistics. Their perceptions show a positive relationship between the two terms, whereby one is the improved version of the other. From the managerial point of view, business analytics is an extension of the popular business intelligence.

Corporations are the main beneficiaries of business analytics because it influences diverse aspects like financing, supply chain management, accounting, and human resource management. Application of business analytics in the supply chain, also referred to as supply chain analytics, improves operational efficiency and effectiveness. Small and large corporations work with a chain of suppliers, making it difficult to determine the best vendors. Using business analytics to analyze the dealers' list is indispensable in the business sector to get the most suitable based on diverse assessment variables. According to Chae and Olson (2013), supply chain analytics uses information and technology to improve value chains' capabilities. The process results in productive and reliable supply chains that enable businesses to withstand market competition pressure (Yesufu, 2021).

3   Methodology

Business Analytics is a relatively new and complicated concept in the ever-changing environment. As a result, people often misunderstand and lack clarity about the various aspects of Business Analytics. Yin and Fernandez (2020) argue that Business Analytics's misinterpretations and misconceptions occur because of its newness in the global market. This paper involves a systematic review of several available and credible literature to clear unclear aspects and misunderstandings. The primary purpose is to identify, assess, analyze, and synthesize the data available from past research. The chapter uses findings on different submissions to define, examine the relationship between Business Analytics and other fields, and assess its functionality in different areas. Business analytics is essential for guiding organizations to develop effective and workable strategic plans that fit market conditions. For example, companies use consumer preference data to determine buyers' perceptions regarding a given product or service (Yin & Fernandez, 2020). The management then transforms this data into useful information that can support the decision-making processes. Business analytics empowers different industries to leverage manufacturing, production, marketing, and distribution efficiency. A company like Toyota, among other automotive makers, owes its success to the utilization of business analytics. The firm has a continuous improvement strategy that depends on enormous data to make strategic improvements.

3.1   Information Sources and Selection Approach

As mentioned previously, the Business Analytics field has limited literature, probably because of the developing discipline's novelty (Yin & Fernandez, 2020). It is worth noting the importance of relying on and utilizing credible and recognized information sources to make valid conclusions about a given topic. What is more, not every document of research meets the quality standards or is credible enough to make valid conclusions according to this study's purpose. For this reason, this paper obtains data from two databases, including Scopus databases and Scopus Journals. The sources' selection is because they are recognized widely and have high-quality publications, particularly the peer-reviewed articles. To settle on the best or the highest quality and reliable data, the systematic review had guidelines that assisted in the selection process.

According to Okoli (2015), a systematic analysis' success depends on the establishment and implementation of rules to screen various studies before selecting. The first qualification was that each paper should have business analytics as a theme and have its main focus on the multiple aspects of the field. The second factor was that the studies must use English to make the analysis easier because this submission uses a similar language. Data or information changes rapidly as a result of dynamics in the world. For instance, technological advancement that increases researchers' access to improved data affects the quality of information concerning time. This factor made the third rule that demanded literature between 2010 and 2020 because it is considerably updated. The paper also considered the reviews about a given article and used the number of citations as a key performance indicator. The assumption was that there is a direct proportionality between the number of times other researchers cited an article with its quality. Lastly, reading through all the research discussing business analytics is a complex, tedious, and time-intensive process. To follow the methodology defined, the selection processes concentrate on papers' abstracts and conclusions to determine whether it fits this journal's research purpose. Most importantly, the selected article should be long enough to cover the BA's diverse and relevant content.

3.2 The Selection Process

The data available is overwhelming and can be confusing when choosing the right one for analytical purposes. There were one hundred samples of Scopus databases, and Scopus Journals selected, awaiting an evaluation. The majority were written in English, and only twenty were in other languages. Out of the remaining samples, fifty had many citations according to the demand, but fifteen were older than 2010. Moreover, ten out of the remaining thirty-five were not long enough, and others had irrelevant information even if the titles were about business analytics. After filtering all the documents using the predetermined rules and assumptions, this paper systematically reviewed fifteen credible and recognized literature.

4   Results

4.1   Definitions of Business Analytics

In the study by Yin and Fernandez (2020), Business analytics is a broad concept covering numerous difficulties and solutions, including planning the resource capacities, workforce, predetermining demand and conditioning, optimization, analyzing consumers and products, modeling sales force, and projecting revenue generation among others. This definition or view outlines the endless reasons businesses need to integrate Business analytics in the operations and make it an integral element of their cultures. It is worth noting that the highlighted components of Business analytics work hand in hand toward an organization's success (Jahantigh et al., 2019). For instance, when determining the expected demand, the management should have sufficient employees and resources to manufacture, market, and deliver goods and services to the target buyers. The decisions should also come up with ways to optimize revenue without compromising all other organizational functions (Yesufu, 2020).

4.2   Relationship of Business Analytics with Business and Other Fields

Business analytics also interrelates with other disciplines. In their study, Varshney & Mojsilovic (2011) find a positive correlation between BA and management science. They argue that Business analytics is applicable in management science through computer science, signal processing, statistics, and operational research. The main difference between management science and Business analytics is probably the setting that each applies. In another study by Parks and Thambusamy (2017), Business analytics integrates useful and measurable knowledge to tactical and strategic corporate goals using databases to make decisions. In this case, data is an assortment of certain observations or factual objectives. On the other hand, descriptive assessment acknowledges the trends and events happening in the investment environment to decide on a given business course. The process can involve a detailed analysis of the market situation like changes in consumer's tastes and preferences and competition to develop competitive strategies (Yin & Fernandez, 2020). Some strategies can include cost leadership, market expansion, and diversification to satisfy the immediate customer needs, increase the market share, and become competitive to withstand intense rivalry. Companies that seek smarter decisions and improved business value can use the three categories of business analytics. Other definitions of business analytics include the process of employing big data evaluation in an investment. Moreover, Shen and Tzeng (2016) argue that Business analytics depends on several other techniques such as data warehousing, mining, modeling, statistical, visualization, quantitative analysis, and other technologies.

Business analytics is more popular in today's business environment because it answers questions concerning optimization and forecasting due to its machine learning techniques. Moreover, Chen et al. (2012) argue that the terms business intelligence and business analytics emerged in literature in the 1990s and 2001, respectively. The latter became popular after 2007 and is currently the most sort after or researched topic today. The study also notes a significant difference between the volumes of information related to the two terms. For example, the numbers of publications about business intelligence by the end of 2012 were ten times more than business analytics, but the researches about Business intelligence are increasing rapidly compared to business intelligence publications. According to Yin and Fernandez (2020), Popovi and colleagues (2012) developed a successful business intelligence system model. The model consists of the main dimensions, such as the culture of making analytical decisions, the maturity of a particular business intelligence system, quality of the information content, and application of informed decision-making processes. The maturity of business analytics describes the state of development of the analytical methods available in a given firm. The quality of information content means the output quality during the collection, evaluation, and data analysis. Arguably, high-quality information content allows companies to make reliable, strong, and valid conclusions concerning a specific or desired business course. Popovič et al. (2012) use the term quality of information access to represent the interactivity, customized abilities, and bandwidth that a certain Business analytics offers to its stakeholders. The success of Business analytics also depends on the organizational culture. Companies that rely on making analytical and thought-through decisions are the best users of Business analytics.

Lastly, the author uses the term informed decision-making to describe a situation whereby an organization applies and transmits the information acquired into strategic decision-making activities. In a study to test the model, Popovič et al. (2012) used 181 samples of data collected from different institutions (Yin & Fernandez, 2020). The analysis discovered that the quality of information content was preferable when making a decision instead of access to information quality. The results also showed that BA's maturity has a significant influence on access to high-quality data, which, in return, affects the quality of decisions that stakeholders make. The study showed a positive correlation between informed decision-making and analysis-based decision-making culture. In this case, companies with a strong culture improved the capacities or dependency on information-based decision-making. The main argument is that the management must place a good culture that allows or increases reliance on analyzed information to develop quality plans and strategies. Most importantly, Yin and Fernandez (2020) emphasize that there is a direct effect of quality content information on BA's successful implementation in different organizations.

4.3   The Functionality of Business Analytics

Acquiring Business analytics capabilities is one of the most critical objectives of many businesses. According to Bayrak (2015), Business analytics is among the top five items that people search on the Internet, and the recent publications about the topic are becoming hits. Thus, companies like IBM, Deloitte Consulting, and Accenture have established analytical practices and centers. Business analytics is growing enormously and rapidly, and the increasing demand for structured data is influencing development. What is more, no other emerging trend in the business environment can have much influence as Business analytics. The study by Shen and Tzeng (2016) indicates that Business analytics has four fundamental functions. The first function is assisting small and huge investments the make the decision making the process more objective than just deciding using assumptions and experiences. Arguably, some businesses fail because the management makes decisions that lack objectivity, particularly when using the past to predict the future. The Business analytics's second function is enhancing and increasing the efficiency or effectiveness of the decision-making process (Yin & Fernandez, 2020). Making good decisions need all the parties involved to have sufficient information to consider many alternatives before arriving at a given conclusion. Accessing large information can be labor and time-intensive, which, in return, can reduce the effectiveness of the decision-making procedure. Business analytics offers a reliable solution to the decision-making challenge by transforming data into information quickly and reliably (Yesufu, 2021).

Business analytics's third function is providing better analytics of consumers and markets, which helps businesses modify their services to buyers and the quality of products according to the market demands. A company can use the consumers' tastes and preferences or their concerns or responses to its quality standards to make the necessary adjustments where necessary (Yin & Fernandez, 2020). The market statistics about competition, strategies used by the rival companies, and new entrants' rates can help a company analyze its business and corporate level approaches to determine their strategic fit with the market situation. The fourth function of BA is enabling investors to understand the external environment, including its main stakeholders. The other three purposes focus mainly on the decision-making process (Calof et al., 2015). In contrast, the last functions consider external environment conditions and key stakeholders to assess a company's chances to succeed or meet its objectives.

Apart from business analytics and business intelligence, another important term in the sector is referred to as business intelligence and analytics. The concept is becoming more common in modern research than before, whereby scholars attempt to merge business intelligence and Business intelligence. Individuals who use the term illustrate the correlation between business intelligence and BA, particularly their interdependency. Nevertheless, Yin and Fernandez (2020) argue that business intelligence and analytics are not necessary. The concept emerged in 2020 before people recognized the importance of business analytics due to its unpopularity. The main use of business intelligence is presenting the historical data to investors to make appropriate growth and development strategies. Business intelligence achieves this function through descriptive evaluation, which is an integral component of business analytics. This section illustrates that business analytics and business intelligence are almost similar, particularly in terms of functionality. It is worth noting the positive correlation between the two words, whereby they work interchangeably to improve corporate competitiveness.

4.4 Application of Business Analytics in Industry

The literature review section identified multiple purposes of business analytics. The main function is to assist businesses to make plausible decisions to enhance operation efficiency, maximize output, and optimize revenues and profitability. Companies also use business analytics to scan the external environment to determine trends, which helps them adjust their strategies to adapt to changes and utilize the available opportunities. In other words, the application of business analytics in the company and industry level is comprehensive (Yin & Fernandez, 2020). People consider sectors such as public security, healthcare, and politics as nonbusiness fields. However, Chen et al. (2012) outline the importance of evaluating the sectors from a business perspective and use business analytics from a broader point of view. E-government and politics use some techniques of business analytics to analyze public opinions and social networks. Politicians assess the chances of winning an election using business analytics techniques such as information mining and prescriptive analysis. The government also uses business analytics to gather enough citizens' opinions before carrying out various public development projects. Furthermore, in public security and safety, the administration leverages business analytics to analyze criminal networks, cyberattacks, sentiment and effect, spatial-temporal, and criminal association, among other issues that bother public safety. According to Calof et al. (2015), developed nations integrate business analytics to assess, monitor, and anticipate the occurrence of famine and epidemics. Disasters such as hurricanes, earthquakes, infectious diseases, famine, and storms affect human well-being, deaths, and economic downturns. Business analytics also assists the government and other institutions to make policies using web data support. Most importantly, business analytics enable forecasting and estimation of outcomes of policy implementation.

Healthcare integrates business analytics in three main aspects, including electronic health records (EHRs), genomics data, and social media information. EHRs are modern system that enables health practitioners to store and retrieve enormous patients' well-being data. This information can include health history, insurance covers, and the last clinic's detail. According to Chen et al. (2012), many hospitals worldwide use electronic data recording systems because of the perceived potential to improve healthcare. Other forms of healthcare data include information from pharmaceutical firms, health facilities, and insurance companies. Business analytics in healthcare facilitates data mining and translating it to meet patients' well-being demands and other stakeholders in the industry (Yin & Fernandez, 2020). Nevertheless, caregivers face three primary challenges when implementing business analytics in the sector.

Firstly, the poor relationship between healthcare organizations makes it difficult to extract or access certain data concerning patients. Some institutions might be unwilling to disclose certain information, even the most needed, which affects the industry's performance. Secondly, many states do not define the standard of healthcare data (Ward et al., 2014). As a result, caregivers can use business analytics to analyze inaccurate information and make invalid conclusions. Thirdly, the medical field lacks competent analytical professionals to implement business analytics successfully. It is worth noting that BA's integration to assemble patients' information facilitates policy-making, enhances the distribution of medical resources, and improves healthcare delivery efficiency. Business analytics also apply to the sports sector. According to Troilo et al. (2016), stakeholders in the industry use sports analytics to maximize profits. The gaming companies, particularly sports betting organizations, use business analytics to predict and prescribe their future performance. Club managers are using business analytics to enhance the productivity of their players. Nevertheless, the majority perceive business analytics in terms of increased revenues.

4.5 Training and Education Situation

Business needs skilled professionals that can understand their needs and interpret different information to enhance their competitiveness. For instance, successful organizations should have competent analytical skills for big data to make informed decisions. The field is relatively new and rapidly developing, but the main challenge is the lack of sufficient experts. Therefore, it is beneficial to understand the current training and education situation.

First, many universities have business analytics programs offered in the business departments. It is worth noting that the departments within the business schools provide industrial engineering, statistics, and computer science above the common courses like marketing, human resource management, and information systems, among others. Second, companies offer training programs in BA (Sharda et al., 2013). The social resource management department's role is equipping workers with useful skills to enhance individual and overall production. Organizations that offer excellent training and development programs have low employee turnover and optimized labor output. Business analytics uses several aspects of data sciences, like algorithms and statistics. The use of computers, science, management systems, and statistics and information systems facilitates business analytics courses (Chen et al., 2012). Business analytics programs' arrangement is relatively independent, although all courses have a theory, applied, and practical sessions. Business analytics classes also emphasize the breadth of information instead of depth. In this case, the course enables learners to use different levels of analytical techniques to address many decision-making programs. Business analytics programs' primary purpose is equipping professionals with solid skills and knowledge to integrate technology into business. They enable learners to convert structured and unstructured data into useful information to support decision-making. Business analytics is a business-oriented course and should incorporate a broad knowledge of business concepts such as management, communication, and negotiation skills.

5 Discussion

5.1 Principal Findings

Business intelligence is most common in information technology settings, while business analytics is popular in the investment community. The integration of information technology in the corporate environment influences the creation of enormous and complicated datasets for different business functions. One of the biggest challenges for many organizations today is recognizing the business purpose and formulating decisions based on such datasets. According to Bayrak (2015), information technology describes organization dependency on large data as Big Data, which signifies the large size and complexity of datasets. Furthermore, the traditional investment model lacks sufficient capacity to evaluate and transform enormous datasets into useful information during decision-making processes (Yin & Fernandez, 2020). Business analytics is the new and most reliable solution to this challenge. The concept allows investments to analyze voluminous databases and interfaces statistically and mathematically to evaluate past and present situations and predict the future or the way forward. Business analytics' definition indicates that the concept has three main levels or categories, including descriptive, prescriptive, and predictive. These levels suggest that there is a significant difference between business analytics and business intelligence. Business intelligence mainly uses data visualization as the representative technology (Sharda et al. (2013). The author also argues that business intelligence depends much on descriptive analysis, which scans the environment to determine the development strategies. It is worth noting that business intelligence applies some concepts and components of business analytics, including its levels. Varshney and Mojsilovic (2011) have similar opinions and suggest that business intelligence is useful during data reporting cases. However, the scholars outline that lack of analytical modeling and endorsements offered by algorithms of decision-making limits business intelligence.

Business analytics also interrelates with other disciplines. In their study, Varshney & Mojsilovic (2011) find a positive correlation between BA and management science. They argue that Business analytics is applicable in management science through computer science, signal processing, statistics, and operational research. The main difference between management science and Business analytics is probably the setting that each applies. In another study by Parks and Thambusamy

(2017), Business analytics integrates useful and measurable knowledge to tactical and strategic corporate goals using databases to make decisions. In this case, data is an assortment of certain observations or factual objectives.

Leveraging business analytics in investment is all about maximized utilization of external and internal data to enhance its efficiency and performance. A business achieves such an improvement because of the previously discussed categories of analytics: descriptive, prescriptive, and predictive. The predictive analysis enables companies to predict the future, such as the impacts of policy changes or the strategies adopted by the market rivals. According to Appelbaum et al. (2017), the prescriptive analysis provides sufficient and dependable facts and recommendations during the decision-making process through the influence of machine learning technologies.

Lastly, the descriptive form of analytics helps small and large corporations access full updates of the present and past situations. For example, the company's performance in the past and today help decision makers or management determines financial and market positions. This information is crucial in strategic planning and understanding the resource allocation approaches to improve situations. The insights acquired through the three categories of analytics are incredible and vital for companies that seek growth, development, productivity, profitability, and competitiveness. According to Parks and Thambusamy (2017), many kinds of research in the field mainly focus on integrating Business analytics in the marketing sector. In this case, corporations use the external and internal data to gather and transform the information acquired to improve its market performance, including expanding the market share, influencing consumer satisfaction and loyalty, and becoming competitive. It is worth noting that data science is not limited to algorithms and statistical dimensions. In this case, data science comprises an intersection of statistics, programing, and data analytics. In other words, business analytics results from data science. One can evaluate the difference between the two concepts in terms of big data. According to Parks and Thambusamy (2017), big data or business analytics has four dimensions: velocity, variety, volume, and veracity. Volume represents the magnitude or the amount of data. Arguably, the volume of data assessable for many companies is growing rapidly due to technological advancement and increasing awareness about the importance of holding essential data. Data scientists must process enormous data acquired from external and internal sources. This large amount of information has technical challenges that data scientists must address to make valid conclusions. Retrieving useful information from voluminous data and utilizing it in decision-making is more challenging as data becomes more complex and large.

Parks and Thambusamy (2017) define variety as various types of data, including documents, hierarchical, tabular, email, audio, financial transactions, stock ticker, still images, and video. Yin and Fernandez (2020) argue that companies today combine structured and unstructured data to acquire a complete outlook of operations and consumers. Moreover, they use raw data like events, images, social media, emails, external feeds, sensors, and log data to have a good picture of the market or the external environment. Velocity is the rate of generating and the speed of analyzing and transforming data into useful information. Handling data quickly and efficiently is one of the main problems that data scientists face. Veracity refers to data uncertainty or abnormality, noise, and biases. Unlike business analytics, data science deals with incompleteness, latency, inconsistency, deception, and model approximation. Business analytics influences decision-making, which leads to better business performances. According to Shen and Tzeng (2016), the retailing sector's success also depends on the proper implementation of business analytics. Sole proprietors and retailers generate consumer shopping and preference information, a process commonly known as basket analysis. Griva et al. (2016) also describe the process as retail business analytics. The sector can also collect customer visit information using the mining techniques of business analytics. This data is critical because it guides, retailers on product and service customization. For example, a company like Amazon, the leading online retailer, mines consumer data to develop strategies that satisfy their immediate needs. Amazon understands that buyers like convenience, quality, and value meet the demands using its consumer-centric, differentiation, and cost leadership strategies.

Business analytics is useful for different aspects of business like accounting, supply chain, and human resource management. Investors integrate the descriptive level of business analytics for financial reporting to show the positions of a company. Investors can combine the data obtained using descriptive techniques with machine learning approaches to forecast their future enterprises. This process is known as predictive analytics, which is an integral component of business analytics. According to Appelbaum et al. (2017), corporations can utilize business analytics's prescriptive level to make important recommendations to improve future performance. Business analytics assists in generating financial time series that enable decision-making, particularly on matters that involve finances. Globalization and internationalization of trade influence a continuous change in how various industries perform (Yesufu, 2018). For instance, increased consumption of advanced technology, particularly the Internet of Things, is interfering with brick-and-mortar stores' normal operations because many consumers are shifting to online shopping. Firms also face a challenge to satisfy unpredictable and always changing consumer behaviors, which affect their competitiveness and survival in an intense rivalry market. Similarly, the introduction of business analytics in different industries is instrumental in improving capacities to achieve their primary purposes. According to Chen et al. (2012), the improved version of business analytics depends on unstructured and web data. The authors point out a significant difference between unstructured data that is based on the website and traditional transactional data. The latter has a good structure that makes it easier for enterprises to understand the main consumers' perceptions and interests.

Training business analytics professionals at the graduate level are ideal using the following alternatives. First, learning institutions should start a master's degree in business analytics to make it affordable to many students. Second, universities should offer business analytics certification programs for people working in business or information and technology through on-site and online sessions. Third, tertiary institutions should modify the current master's courses in IT and introduce business analytics programs (Yin & Fernandez, 2020). Business analytics programs differ from pure data sciences, which is why universities offer them in business schools.

5.2   Strengths and Limitations of Study

The major strength of this study is using more updated and recent papers. The research and education field is changing rapidly, which limits dependency on older researches. This paper utilized studies between 2010 and 2020 to get current information. The methodology and selection processes of the journal papers utilized some rules such as the citation index to assess credibility and quality. The information on research journal is applicable to business and industrial sectors in diverse aspects like accounting and supply chain, finance, and human resource management. It also equips learners with critical business analytics tools essential for their future career. The paper's content is useful in the private and public sectors.

5.3   Practical Limitations

This paper depended entirely on other researches. The methodology filtered many peer-reviewed articles using assumptions and rules to get the most credible. Given the time limitation and tedious task of reading through every paper, the section criteria focused on abstracts and conclusions only. This submission lacks sufficient proof that the fifteen researches selected were the best. Working within a limited duration was another practical challenge of this information. However, the paper maintained a detailed evaluation of the identified literature about the topic with an international focus, which enhances its quality.

5.4   Research Limitations

This research depended on secondary data. The main challenge is determining the accuracy and feasibility of findings in the papers analyzed. For instance, the submission suggests that many businesses use business analytics to enhance the decision-making process. However, the study does not utilize statistical facts and figures such as percentages and bar graphs to make valid comparisons and conclusions. There is a statistical knowledge gap in this study that requires a quantitative research to determine rates of business analytics integration in investments and its financial benefits.

6   Conclusions

Business analytics is a relatively new but rapidly growing concept in the global environment. Many companies now recognize the importance of integrating Business analytics to address challenges, optimizing data value, and forecasting future performance. More and more enterprises are increasing their dependency on business analytics techniques to influence decision-making processes. Given the rapid globalization, continued changes in the external environment, technological advancement, changing consumers' behavior, and intense market competition, investors must make plausible decisions to satisfy market demands and enhance competitiveness. Advanced business analytics help investors recognize the importance of data and develop strategies based on statistics. The literature about business analytics is limited, perhaps because of the novelty of the discipline. This study reviewed some credible studies systematically to understand the meaning, application, and the relationships between business analytics and other fields. Business analytics provides techniques to collect, analyze, and transform datasets from different business databases into useful decision-making insights. Every business that seeks success in the global market must integrate business analytics concepts and make it an integral part of their culture. Universities and companies should muscle up the provision of business analytics education and training to increase business analytics professionals.

References

Appelbaum, D., Kogan, A., Vasarhelyi, M., & Yan, Z. (2017). Impact of business analytics and enterprise systems on managerial accounting. *International Journal of Accounting Information Systems, 25*, 29–44.

Bayrak, T. (2015). A review of business analytics: A business enabler or another passing fad. *Procedia-Social and Behavioral Sciences, 195*, 230–239.

Brock, V. F., & Khan, H. U. (2017). Are enterprises ready for big data analytics? A survey-based approach. *International Journal of Business Information Systems, 25*(2), 256–277.

Calof, J., Richards, G., & Smith, J. (2015). Foresight, competitive intelligence and business analytics—Tools for making industrial programmes more efficient. *Journal of the National Research University Higher School of Economics., 9*(1), 68–81.

Chae, B., & Olson, D. (2013). Business analytics for supply chain: A dynamic-capabilities framework. *International Journal of Information Technology & Decision Making, 12*, 9–26. https://doi.org/10.1142/S0219622013500016

Chen, H., Chiang, R. H., & Storey, V. C. (2012). Business intelligence and analytics: From big data to big impact. *MIS Quarterly*, 1165–1188.

El Alfy, S., Gómez, J. M., & Dani, A. (2019). Exploring the benefits and challenges of learning analytics in higher education institutions: A systematic literature review. *Information Discovery and Delivery., 47*(1), 25–34.

Gorman, M. F., & Klimberg, R. K. (2014). Benchmarking academic programs in business analytics. *Interfaces, 44*(3), 329–341.

Griva, A., Bardaki, C., Pramatari, K., & Papakiriakopoulos, D. (2016). Retail business analytics: Customer visit segmentation using market basket data. *Expert Systems with Applications, 100*, 1–16.

Jahantigh, F. F., Habibi, A., & Sarafrazi, A. (2019). A conceptual framework for business intelligence critical success factors. *International Journal of Business Information Systems, 30*(1), 109–123.

Mashingaidze, K., & Backhouse, J. (2017). The relationships between definitions of big data, business intelligence and business analytics: A literature review. *International Journal of Business Information Systems, 26*(4), 488–505.

Okoli, C. (2015). A guide to conducting a standalone systematic literature review. *Communications of the Association for Information Systems, 37*(1), 43.

Parks, R., & Thambusamy, R. (2017). Understanding business analytics success and impact: A qualitative study. *Information Systems Education Journal, 15*(6), 43.

Popovič, A., Hackney, R., Coelho, P. S., & Jaklič, J. (2012). Towards business intelligence systems success: Effects of maturity and culture on analytical decision making. *Decision Support Systems, 54*(1), 729–739.

Priya, L. K., Devi, M. K., & Nagarajan, S. (2017). Data analytics: Feature extraction for application with small sample in classification algorithms. *International Journal of Business Information Systems, 26*(3), 378–401.

Sharda, R., Asamoah, D. A., & Ponna, N. (2013). Research and pedagogy in business analytics: Opportunities and illustrative examples. *Journal of Computing and Information Technology, 21*(3), 171–183.

Shen, K. Y., & Tzeng, G. H. (2016). Contextual improvement planning by fuzzy-rough machine learning: A novel bipolar approach for business analytics. *International Journal of Fuzzy Systems, 18*(6), 940–955.

Sivarajah, U., Kamal, M. M., Irani, Z., & Weerakkody, V. (2017). Critical analysis of big data challenges and analytical methods. *Journal of Business Research, 70*, 263–286.

Troilo, M., Bouchet, A., Urban, T. L., & Sutton, W. A. (2016). Perception, reality, and the adoption of business analytics: Evidence from north American professional sport organizations. *Omega, 59*, 72–83.

Vanani, I. R., & Jalali, S. M. J. (2018). A comparative analysis of emerging scientific themes in business analytics. *International Journal of Business Information Systems, 29*(2), 183–206.

Varshney, K. R., & Mojsilović, A. (2011). Business analytics based on financial time series. *IEEE Signal Processing Magazine, 28*(5), 83–93.

Ward, M. J., Marsolo, K. A., & Froehle, C. M. (2014). Applications of business analytics in healthcare. *Business Horizons, 57*(5), 571–582.

Wixom, B. H., Yen, B., & Relich, M. (2013). Maximizing value from business analytics. *MIS Quarterly Executive, 12*(2).

Yesufu, L. O. (2018). Motives and measures of higher education internationalisation: A case study of a Canadian university. *International Journal of Higher Education, 7*(2), 155–168.

Yesufu, L. O. (2020). The impact of employee type, professional experience and academic discipline on the psychological contract of academics. *International Journal of Management in Education, 14*(3), 311–329.

Yesufu, L. O. (2021). Predictive learning analytics in higher education. In *Data analytics in marketing, entrepreneurship, and innovation* (p. 151).

Yin, J., & Fernandez, V. (2020). A systematic review on business analytics. *Journal of Industrial Engineering and Management, 13*(2), 283–295.

Green Purchasing Behavior on Social Media: A Goal-Framing Theory Perspective

Kashif Farhat, Wajeeha Aslam, and Shahira El Alfy

1 Introduction

Environmental issues have become increasingly prominent, posing a severe threat to human growth and a source of public concern (Dangelico & Vocalelli, 2017; Szabo & Webster, 2021). Following the realization of the immense harm caused by the ecological catastrophe, the government, businesses, and social media have adopted a variety of steps to promote environmental conservation in recent years (ElHaffar et al., 2020; Sangroya et al., 2020; Yang et al., 2020). For instance, a significant amount of information on the environmental problem was present to various media outlets to promote the concept of sustainable development and encourage people to engage in pro-environmental activities in their everyday lives (Ivanova et al., 2019).

Personal experience with ecological crisis and external rewards, such as media information, may add value to residents' positive attitudes (willingness) toward environmentally sustainable products. However, both professionals and studies have discovered that customers' positive green purchasing attitude does not always lead to more purchase decisions (Nguyen et al., 2019). This eco-friendly motivation–behavior gap has become the primary impediment to the healthy growth of the green consumption sector (Nguyen et al., 2019). How to bridge this chasm has become a pressing concern for green professionals and academic scholars alike.

K. Farhat
Department of Marketing, Mohammad Ali Jinnah University, Karachi, Pakistan

W. Aslam
Department of Business Administration, IQRA University, Karachi, Pakistan

S. El Alfy (✉)
Faculty of Business, Higher Colleges of Technology, Abu Dhabi, UAE
e-mail: selalfy@hct.ac.ae

The past theoretical literature on the development process of green purchasing intention and behavior has included Theory of Consumption Value (Sheth et al., 1991), Theory of Planned Behavior (Ajzen, 1991), and Value-Belief-Norm Theory (Stern et al., 1999), focusing on theories: Value Theory (Schwartz, 1994) and Norm Activation Theory. Even though some of the past scholarly work defines the green motivation-behavior gap, they do not provide a complete and adequate understanding of why there is a gap. Recently, researchers have begun to unearth the numerous reasons for the motivation-behavior gap by employing diverse study frameworks. Most researchers agree that the separation of egoistic and altruistic appeals is the main factor contributing to the gap (ElHaffar et al., 2020; Ling & Xu, 2020). Furthermore, green consumption bears the property of social desirability (Rahman et al., 2020); therefore, social desirability bias is a probable cause of this disparity (Shaw et al., 2016). As a result, these viewpoints can potentially undermine the impact of social media persuasion on green purchasing behavior. In light of this, it is vital to better understand the shaping power of social media persuasion and the many mediating or moderating elements in constructing a green consumer society. In this study, we classify social media purchasing as the shaping power to the degree to which media can strongly impact individuals' cognition of pro-environmental behaviors. Such classification would create, reshape, and change people's pro-environmental psychology and, in turn, behaviors anchored on the S-O-R model (Mehrabian & Russell, 1974).

Previous literature has studied and described the barriers to sustainable consumption using an unlimited moral hypothesis or an unbounded self-serving hypothesis (e.g., Wang et al., 2019). Furthermore, some academics have focused on the trade-off and synergy between egoistic and altruistic appeals (e.g., Yang & Zhang, 2020); however, a solid theoretical framework to systematically examine how to bridge this gap is required. The purpose of this research is to fill this void by presenting a new framework that integrates goal-framing theory in the context of green purchasing behavior. This theory is consistent with the theories of bounded self-serving and bounded ethicality, taking into account the dual appeals (Tang et al., 2020). Furthermore, according to behaviorism's learning theory, media, as a key external stimuli factor, may significantly influence the construction of an individual's goal frames (Donmez-Turan & Kiliclar, 2021). Various situations might trigger distinct goal frames, which can have varied consequences on an individual's pro-environmental actions. While Chakraborty et al. (2017) detailed the influence process of three-goal frames on an individual's pro-environmental behavior; they forgot to include the method by which goal frames are formed (motivations). As a result, this study investigated the mechanism of social media persuasion impact on green purchase by molding the three-goal frames (behavioral motivation) and compares the intensity of the three mediation channels.

Furthermore, the study investigates the mechanism of the green motivation-behavior gap from the standpoint of dual appeals, consequently providing viable rectification measures. Lastly, the influence of media "fear appeal" has long been questioned (Chen & Yang, 2019; Mohd Suki & Mohd Suki, 2019). Earlier, it was argued that low fear appeal was more convincing than high fear appeal due to the "reversion effect" (i.e., the audience's favorable attitude returned to the prior indifference when confronted with high fear) (Rogers & Mewborn, 1976).

On the other hand, recent research has discovered that "fear appeal" is positively related to customers' green purchase behavior (Lee et al., 2017; Mohd Suki & Mohd Suki, 2019). This study includes the perceived severity of environmental concerns (a measure of fear appeal) (PSEP) as a potential moderator. Such inclusion helps examine its other possible formation processes of "reversion impact" on green purchasing behavior to analyze the efficacy of the "fear appeal" strategy in the context of sustainable consumption. This work sought to make some theoretical and empirical implications for expanding green purchase behavior.

2 Literature Review

Lindenberg and Steg (2007) introduced the goal-framing theory to integrate and systematically explain the role of motivation in the development of an individual's pro-environmental conduct. According to this theory, goals are significant considerations in steering behaviors (Lindenberg & Steg, 2013; Steg et al., 2014). Each goal frame correlates to a behavioral motive (Lindenberg & Steg, 2007; Tang et al., 2020; Wang et al., 2019). Lindenberg and Steg (2007) subdivided the goals into three components: gain goal frames (GGF), hedonic goal frames (HGF), and normative goal frames (NGF). Individuals' everyday activities under the GGF orientation follow the concept of benefit maximization, seeking their objectives and resources to survive, and hoping to gain individual interests or enhance the utilization rate of resources at the lowest possible cost. People who focus on joyful experiences and good emotions received during and after acts, and the goal of behavior is to pursue positive subjective feelings while avoiding painful feelings, according to the HGF.

In contrast to the GGF and HGF, the NGF contends that people are attentive to public interests intentionally and independently, motivated by altruistic incentives while adhering to societal norms and moral codes, and are eager to take on social obligations. According to Lindenberg and Steg (2007), human conduct is the product of these many incentives. Whereas the GGF may be the primary motivator for conduct, additional objectives will either reinforce or diminish the impact of present goals, altering behavioral style in tandem (Tang et al., 2020).

2.1 Social Media Persuasion and Goal-Fain Frame Theory

Generally, media persuasion affects the behavioral patterns of individuals. In other words, media persuades individuals to think and act in a certain way in a particular situation. In environmental issues context, social media provides customers with relevant information, which often enhances the environmental concern and positive attitude toward the environment (Ghermandi & Sinclair, 2019; Zheng et al., 2019). Essentially, goals' frame' the process of information people adopt and follow through on it. Usually, many goals are present at any given moment, which may be compatible, and strengthen focal goal persuasion in the presence of secondary goals. As unsustainable energy, food, and natural resource consumption has caused a range of widespread social and environmental problems, sustainable products are developed and provided in consumer markets (Singh, 2019). However, the slow adoption of sustainable products has been a concern for marketers and governments alike (Cerri et al., 2018; Jansson et al., 2017). While we understand the media has a role in adopting sustainable products, i.e., green purchase behavior (Yang et al., 2020), little is known how social media drives green purchase behavior. Framing theory indicates that "framing," or how an object is presented to individuals, determines how they will process the information and make future choices. For instance, social media persuades individuals to adopt altruism in life (Mahmood et al., 2019). Likewise, green advertising is designed to emphasize persuading green consumption behavior (Bedard & Tolmie, 2018).

Unsustainable or reckless use of food, energy, and natural resources has resulted in a slew of social and environmental issues. In reaction, corporations and governments coordinate to develop and promote sustainable products with less negative environmental consequences. However, consumer acceptance of environmentally friendly items is gradual (Boukid, 2020; Kucharčíková & Mičiak, 2018). According to framing theory, the media concentrates attention on specific occurrences and then frames them into a sphere of meaning. In principle, framing theory proposes that how something is presented to the audience (referred to as "the frame") impacts individuals' decisions about interpreting that information. The media convinces audiences to practise charity in their daily lives and to choose green consumerism. Many businesses use green advertising and place high importance on green marketing methods. Nonetheless, insufficient studies have been inducted to study the efficacy of green appeal in social media commercials. Based on message framing and contextual factors level theory, this study investigates the function of temporal distance in moderating the effects of gain or loss framed messages on consumers' attitudes and buying intentions toward the promoted brand (Cheng & Wu, 2015).

H1 Social media persuasion toward environmental issues has loss framed influence on goal gain frame.

The media frequently plays a crucial role in conveying a key message to customers that affect their collective concern for the environment (Han & Xu, 2020). According to Swenson and Wells (2018), environmental concerns are more susceptible to media coverage than other social issues. Moreover, Hedonic items are purchased for the customer to derive pleasure and satisfaction from the item. Hedonic consumerism, which results from receiving emotional gratification from purchase, takes precedence over genuine necessity.

Using hedonic items may allow individuals to be delighted and enjoy sensory pleasure (Govind et al., 2020), but it may also lead to regret (Baghi & Antonetti, 2017). The media portrays environmental issues so that an individual becomes emotionally moved to purchase a product. With immediate effect, media can modify an individual user's perspective about their environmentally undesirable excessive purchasing habit (Wolstenholme et al., 2020).

H2 Social media persuasion toward environmental issues has a positive influence on Hedonic Gain.

The media's dissemination of environmental crisis information heightens communities' environmental sensitivities (Trivedi et al., 2018). The media influence people and society to shift their attention away from other concerns to environmental ones. The media is skilled at creating scenarios and presenting them so that they capture people's attention rapidly.

According to the Norm Activation Theory (Shin et al., 2018), an individual's knowledge of consequence and attribution of responsibility for non-environmental activities would activate personal norms, resulting in forming a normative gain frame. The inner feelings of a person cause him to be responsive to climatic problems, doing the right things and rejecting activities that damage the environment. The normative purpose of an individual pushes him to pay attention to environmental issues and be more concerned about solving them.

H3 Social media persuasion toward environmental issues has a positive influence on the normative gain frame.

Environmental attitude has been recognized as one of the keys and most critical antecedent variables that impact future purchase intention and environmentally concerned customers' behavioral actions to better understand their behavior (Prete et al., 2017). An individual's internal environmental attitude refers to conduct that evolves into a positive attitude toward an object or issue. Prior research has demonstrated that the media may directly influence consumers' environmental views and behavior on a wide range of issues, including energy, pollution levels, and the repercussions of ecological deterioration (Chan & Lau, 2000; Liu et al., 2021).

H4 Social media persuasion toward environmental issues has a positive influence on inward environmental attitudes.

Milfont (2007) defined environmental attitude as a "psychological propensity" that leads to perceptions or beliefs about the environment. The media has a significant impact on altering people's environmental attitudes toward green products. Outward environmental attitude is described as "attitudes toward the perceived need for social, political, and legal reforms to safeguard the environment" (Leonidou et al., 2010). Anyone who uses social media or follows the news is more likely to become interested in political and social developments and will participate actively in any environment protection behaviors.

H5 Social media persuasion toward environmental issues has a positive influence on outward environmental attitudes.

Green purchase behavior is defined in this study as making ecologically responsible purchases that are recyclable, preserve resources, or improve the environment

(Chen et al., 2020; Gonçalves et al., 2016). Previous research has also discovered a link between goal frames and pro-environmental behavior (Tang et al., 2020). According to Wang et al. (2019), when customers discover that green products may suit their egoistic needs, they are more likely to buy them. Consumers' key incentive for purchasing green items is a utilitarian appeal (Chen, 2013). Wang et al. (2019) also determined that the top three reasons customers choose sustainable goods are safety and health, dependable quality, and cost savings. Chen (2013) verified that green perceived value achieves consumer loyalty to green products through customer pleasure and trust, resulting in repeat purchase behavior.

H6 The goal gain frame has a positive influence on green purchase behavior.

Meneses (2010) and Gonçalves et al. (2016) cite feeling as a factor influencing pro-environmental behaviors. Previous research has looked at the effect of mood and emotional experiences on environmental practices. According to Wang et al. (2017) positive emotional appeals effectively increase customers' propensity to purchase sustainable products. Positive feelings such as appreciation and pride are thought to be major psychological motivators that could lead customers to purchase sustainable products (Tang et al., 2020).

H7 Hedonic gain frame has a positive influence on green purchase behavior.

When consumers are aware of their environmental obligations or recognize the negative implications of their activity, personal norms can be triggered (Ge et al., 2020). According to Hafner et al. (2019), normative information is compelling in motivating an individual's pro-environmental conduct. The more environmental responsibility is perceived, the greater the consumers' care for the ecology (Yang et al., 2020). Similarly, Yarimoglu and Binboga (2019) contend that sustainability issues beneficially impact customers' pro-environmental behaviors.

H8 Normative gain frame has a positive influence on green purchase behavior.

Unlike the inward environment attitude that includes individuals' personal objectives and behaviors, the outward environment attitude involves increased expression of society concerning environmental problems. Environmental protection is a major public concern in modern societies, and politically engaged individuals are more likely to be receptive to and foster positive attitudes toward environmental preservation (Mikael Klintman, 2013). According to research on the outward environment, a person with a high external environment concern will conduct more suitable behaviors to help people and society (Zhang et al., 2018). Working with public groups, for example, urging parties to cooperate for the environment and joining environmental organizations.

H9 Outward environmental concern has a positive influence on green purchase behavior.

According to a notion of planned conduct (Ajzen, 1991). if a person has a positive attitude about something, they are more inclined to act positively toward that item, and vice versa. According to the above statement, if a particular individual is worried about environmental problems, he will most likely be driven to do all necessary to decrease them. He will be sufficiently motivated to acquire and utilize environmentally beneficial and non-harmful items, as well as recycled products. Previous research, specifically (Alwitt & Pitts, 1996), revealed that environmental concern influenced attitudes toward environmentally friendly items. Alwitt and Pitts (1996) also discovered that individuals who are more concerned about the environment are more likely to have a favorable attitude about utilizing eco-friendly parking facilities.

H10 Inward environmental concern has a positive influence on green purchase behavior.

Ecological concern was defined by Zelezny and Schultz (2000) as "an aspect of the system of belief that indicates to certain psychological factors associated to people' proclivity to join pro-environmental action regimes." The ecological concern is regarded as a mental state study variable. This multidimensional construct ranges from low (generic) to high (product) levels, and it differs from its antecedents as well as behavioral effects (Sharma & Bansal, 2013). Perceived climate costs and perceived climate benefits are the most reliable indicators of readiness to act or support climate strategy initiatives (Tobler et al., 2012). There are several empirical studies available that demonstrate a desirable relationship between environmental concern and green buying (Kautish et al., 2019; Wang, 2014) By considering the moderating influence of pro-social status, a complete gap model was proposed for environmentally concerned consumer behavior between willingness to act and actual green purchase behavior (Zabkar & Hosta, 2013). Moreover, particular recycling attitudes are marginally related to a broad environmental concern (Vining & Ebreo, 1992). As a result, the following hypotheses are developed:

H11 Environmental concern moderates the relationship between inward environmental attitude and green purchase behavior.

H12 Environmental concern moderates the relationship between outward environmental attitude and green purchase behavior.

3   Methodology

3.1   Sampling Technique and Research Instrument

Using the non-probability convenience sampling technique, the data was gathered through a self-administered 7-point Likert scale survey questionnaire from the people of Karachi, Pakistan. The questionnaire was composed of two parts. The first part was related to the demographics of the respondents, such as age, gender, education, and income. The second part contained the items of the constructs that were adapted from past studies. The items of media persuasion were adapted from Yang and Zhang (2020). The items of Goal frames (i.e., goal gain frame, hedonic goal frame, and normative goal frame) were adapted from Tang et al. (2020), and green purchasing behavior's items were adapted from Jaiswal and Kant (2018), and Trivedi et al. (2018). While the items of the ecological concern were adapted from Johnson et al. (2004), and the items of inward and outward environment attitudes were adapted from Leonidou et al. (2010). The questionnaire was designed on Google Docs, and the link to the questionnaire was sent to friends, family, and colleagues through social media platforms such as WhatsApp, Facebook, and LinkedIn.

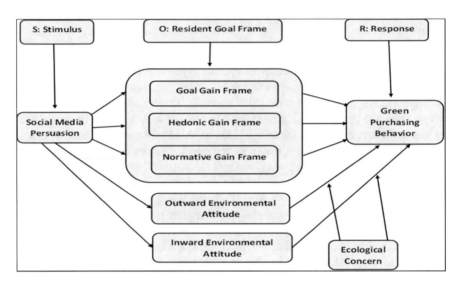

3.2 Pilot Testing and Data Screening

Initially, the data was gathered from 50 respondents to test the reliability of the questionnaire. Upon satisfactory reliability of the pilot test of each construct (Cronbach's Alpha >0.7), the final sample was gathered from 345 respondents which was consequently screened for missing values, out-of-range values, and to eliminate outliers. Altogether, 40 responses were deleted that reduced the sample to 305.

Explaining the demography of the sample, most respondents were male, i.e., 183 respondents. Moreover, majority of the respondents belonged to the age group 28–35 years (i.e., 39.50%), and about 37.7% had bachelor's education, followed by postgraduate education (i.e., 34.7%).

4 Results

4.1 Measurement Model

To test the hypotheses of the study, PLS-SEM technique was employed. Firstly, the measurement model was assessed to check the content validity, construct reliability, convergent validity, and discriminant validity of the data. Cronbach alpha and composite reliability values loaded well above 0.70. Likewise, cross-loadings of constructs were tested to establish discriminant validity of the constructs. Discriminant validity was further established by ensuring that the HTMT values for all constructs were below 0.85. Finally, for convergent validity, the AVE values loaded above 0.50 for all constructs. Thus, the research model of the study exhibited substantial measurement reliability, discriminant validity, and convergent validity.

4.2 Structural Model

To test the hypotheses, bootstrapping was performed using 5000 subsamples as suggested by Hair Jr. et al. (2016) in PLS-SEM. The results of hypotheses testing are mentioned in Table 1. The findings revealed that social media persuasion significantly affects the goal frames. More specifically, social media persuasion largely affects goal gain ($\beta = 0.610$, $p < 0.05$), followed by hedonic gain ($\beta = 0.562$, $p < 0.05$), and normative gain ($\beta = 0.556$, $p < 0.05$). The results also showed that social media persuasion also affects inward environmental attitude ($\beta = 0.393$, $p < 0.05$), and outward environmental attitude ($\beta = 0.369$, $p < 0.05$), respectively.

Table 1 Hypotheses testing

Hypotheses	Coefficient	T Statistics	P Values
Social media persuasion ->goal gain	0.610	14.07	0.000
Social media persuasion ->hedonic gain	0.562	12.468	0.000
Social media persuasion ->inward environmental attitude	0.393	7.354	0.000
Social media persuasion ->normative gain	0.556	11.659	0.000
Social media persuasion ->outward environmental attitude	0.369	6.883	0.000
Goal gain ->green purchasing behavior	0.167	3.258	0.001
Hedonic gain ->green purchasing behavior	0.184	2.604	0.005
Inward environmental attitude ->green purchasing behavior	0.342	4.897	0.000
Normative gain ->green purchasing behavior	0.077	1.142	0.127
Outward environmental attitude ->green purchasing behavior	−0.056	0.911	0.181
Eco*inward ->green purchasing behavior	0.13	2.226	0.013
Eco*outward ->green purchasing behavior	−0.12	1.213	0.113

Moreover, goal gain ($\beta = 0.167$, $p < 0.05$), hedonic gain ($\beta = 0.184$, $p < 0.05$), and inward environmental attitude ($\beta = 0.342$, $p < 0.05$) significantly affects green purchase behavior, whereas; normative gain ($\beta = 0.07$, $p > 0.05$), and outward environmental attitude ($\beta = -0.056$, $p > 0.05$) does not found significant in determining green purchase behavior.

Lastly, the moderating role of environmental concern is found significant for inward environmental attitude and green purchase behavior ($\beta = 0.130$, $p < 0.05$), whereas; it does not moderate the relation between outward environmental attitude and green purchase behavior ($\beta = -0.12$, $p > 0.05$).

5 Conclusion and Implications

The theoretical contribution of the study is first shown in the richness and development of the theoretical framework of green consumption. In reference to the goal-framing theory that considers both egoistic and altruistic appeals, the study provides a theoretical base into the applicability and utility of persuasion theory related to green consumption behaviors, as well as further comprehension of the effectiveness of social media persuasion on broadening green purchase behavior can provide implications for theory for media to optimize their role positioning and strategic plan to protect the environment. Secondly, in the perspective of S-O-R paradigm, this research explores the influence mechanism of social media persuasion on green purchase behaviors. Such influence mechanism is examined by molding customers' three-goal frames and comparing the effect of three mediators from a comprehensive viewpoint that considers customers' attitudinal factors simultaneously. This research gives theoretical proof for social media to maximize their sources of information. Finally, our findings imply that the use of social media persuasion, a "fear appeal" tactic, may have a multifaceted effect on green purchase behaviors, resulting in an "inverse effect" that magnifies the motivation-behavior gap in similar situations.

This study's empirical results have substantial practical implications for managers who encourage green purchases. Research findings show that marketers should minimize highlighting the practical value of green products in marketing efforts. Not only is it easy to fall into the trap of excessive marketing, but this method is also incapable of establishing a long-term mechanism to promote green purchase behaviors. It is notably the case if companies' inventive capacity to produce green goods falls short of consumers' expectations. In this circumstance, consumers' perceptions of greenwashing are readily established, which eventually impedes the healthy growth of the green-consuming industry.

Moreover, when promoting their green products, brands' social media should pay greater attention to the variances in consumers' information processing preferences and processing modalities. Likewise, while performing high-frequency persuasion, social media should maintain relatively uniqueness to decrease the likelihood of demonstrating a defensible mechanism caused by plain repetitive information. Further, social media can use "fear appeal" to boost the moderated function of "environmental concern." This case necessitates that green product managers consider customers' rational and emotional appeals when designing social media marketing content to fully activate individuals' environmental concerns and elicit positive green purchase behaviors.

5.1 Limitations

This study has a few shortcomings that require additional examination and refinement. First, measuring actual green purchase behavior is highly complicated and challenging, owing to negative influence from issues such as social bias. Future research can make even more significant improvements. Researchers, for example, might try identifying actual green customers by studying consumer purchasing circumstances. Second, the findings were drawn from the data collected from a single city. Future research might include employee cross-country inhabitants to compare the findings with this study's findings.

References

Ajzen, I. (1991). The theory of planned behavior. *Organizational Behavior and Human Decision Processes, 50*(2), 179–211. https://doi.org/10.1016/0749-5978(91)90020-T

Alwitt, L. F., & Pitts, R. E. (1996). Predicting purchase intentions for an environmentally sensitive product. *Journal of Consumer Psychology, 5*(1), 49–64. https://doi.org/10.1207/S15327663JCP0501_03

Baghi, I., & Antonetti, P. (2017). High-fit charitable initiatives increase hedonic consumption through guilt reduction. *European Journal of Marketing, 51*(11–12), 2030–2053. https://doi.org/10.1108/EJM-12-2016-0723/FULL/XML

Bedard, S. A. N., & Tolmie, C. R. (2018). Millennials' green consumption behaviour: Exploring the role of social media. *Corporate Social Responsibility and Environmental Management, 25*(6), 1388–1396. https://doi.org/10.1002/CSR.1654

Boukid, F. (2020). Plant-based meat analogues: From niche to mainstream. *European Food Research and Technology, 247*(2), 297–308. https://doi.org/10.1007/S00217-020-03630-9

Cerri, J., Testa, F., & Rizzi, F. (2018). The more I care, the less I will listen to you: How information, environmental concern and ethical production influence consumers' attitudes and the purchasing of sustainable products. *Journal of Cleaner Production, 175*, 343–353. https://doi.org/10.1016/J.JCLEPRO.2017.12.054

Chakraborty, A., Singh, M. P., & Roy, M. (2017). A study of goal frames shaping pro-environmental behaviour in university students. *International Journal of Sustainability in Higher Education, 18*(7), 1291–1310. https://doi.org/10.1108/IJSHE-10-2016-0185/FULL/XML

Chan, R. Y. K., & Lau, L. B. Y. (2000). Antecedents of green purchases: A survey in China. *Journal of Consumer Marketing, 17*(4), 338–357. https://doi.org/10.1108/07363760010335358/FULL/XML

Chen, L., & Yang, X. (2019). Using EPPM to evaluate the effectiveness of fear appeal messages across different media outlets to increase the intention of breast self-examination among

Chinese women. *Health Communication, 34*(11), 1369–1376. https://doi.org/10.1080/10410236.2018.1493416

Chen, Y.-S. (2013). Towards green loyalty: Driving from green perceived value, green satisfaction, and green trust. *Sustainable Development, 21*(5), 294–308. https://doi.org/10.1002/sd.500

Chen, Y.-S., Huang, A.-F., Wang, T.-Y., & Chen, Y.-R. (2020). Greenwash and green purchase behaviour: The mediation of green brand image and green brand loyalty. *Total Quality Management & Business Excellence, 31*(1–2), 194–209. https://doi.org/10.1080/14783363.2018.1426450

Cheng, T.-M., & Wu, H. C. (2015). How do environmental knowledge, environmental sensitivity, and place attachment affect environmentally responsible behavior? An integrated approach for sustainable island tourism. *Journal of Sustainable Tourism, 23*(4), 557–576. https://doi.org/10.1080/09669582.2014.965177

Dangelico, R. M., & Vocalelli, D. (2017). "Green marketing": An analysis of definitions, strategy steps, and tools through a systematic review of the literature. *Journal of Cleaner Production, 165*, 1263–1279. https://doi.org/10.1016/j.jclepro.2017.07.184

Donmez-Turan, A., & Kiliclar, I. E. (2021). The analysis of pro-environmental behaviour based on ecological worldviews, environmental training/ knowledge and goal frames. *Journal of Cleaner Production, 279*, 123518. https://doi.org/10.1016/j.jclepro.2020.123518

ElHaffar, G., Durif, F., & Dubé, L. (2020). Towards closing the attitude-intention-behavior gap in green consumption: A narrative review of the literature and an overview of future research directions. *Journal of Cleaner Production, 275*, 122556. https://doi.org/10.1016/j.jclepro.2020.122556

Ge, W., Sheng, G., & Zhang, H. (2020). How to solve the social norm conflict dilemma of green consumption: The moderating effect of self-affirmation. *Frontiers in Psychology, 11*, 3253. https://doi.org/10.3389/FPSYG.2020.566571/BIBTEX

Ghermandi, A., & Sinclair, M. (2019). Passive crowdsourcing of social media in environmental research: A systematic map. *Global Environmental Change, 55*, 36–47. https://doi.org/10.1016/J.GLOENVCHA.2019.02.003

Gonçalves, H. M., Lourenço, T. F., & Silva, G. M. (2016). Green buying behavior and the theory of consumption values: A fuzzy-set approach. *Journal of Business Research, 69*(4), 1484–1491. https://doi.org/10.1016/j.jbusres.2015.10.129

Govind, R., Garg, N., & Mittal, V. (2020). Weather, affect, and preference for hedonic products: The moderating role of gender. *Journal of Marketing Research, 57*(4), 717–738. https://doi.org/10.1177/0022243720925764

Hafner, B. R., Elmes, D., Read, D., & White, M. P. (2019). Exploring the role of normative, financial and environmental information in promoting uptake of energy efficient technologies. *Journal of Environmental Psychology, 63*, 26–35. https://doi.org/10.1016/J.JENVP.2019.03.004

Han, R., & Xu, J. (2020). A comparative study of the role of interpersonal communication, traditional media and social media in pro-environmental behavior: A China-based study. *International Journal of Environmental Research and Public Health, 17*(6), 1883. https://doi.org/10.3390/IJERPH17061883

Ivanova, O., Flores-Zamora, J., Khelladi, I., & Ivanaj, S. (2019). The generational cohort effect in the context of responsible consumption. *Management Decision, 57*(5), 1162–1183. https://doi.org/10.1108/MD-12-2016-0915

Jaiswal, D., & Kant, R. (2018). Green purchasing behaviour: A conceptual framework and empirical investigation of Indian consumers. *Journal of Retailing and Consumer Services, 41*, 60–69.

Jansson, J., Nordlund, A., & Westin, K. (2017). Examining drivers of sustainable consumption: The influence of norms and opinion leadership on electric vehicle adoption in Sweden. *Journal of Cleaner Production, 154*, 176–187. https://doi.org/10.1016/j.jclepro.2017.03.186

Johnson, C. Y., Bowker, J. M., & Cordell, H. K. (2004). Ethnic variation in environmental belief and behavior: An examination of the new ecological paradigm in a social psychological context. *Environment and Behavior, 36*(2), 157–186.

Kautish, P., Paul, J., & Sharma, R. (2019). The moderating influence of environmental consciousness and recycling intentions on green purchase behavior. *Journal of Cleaner Production, 228*, 1425–1436. https://doi.org/10.1016/j.jclepro.2019.04.389

Kucharčíková, A., & Mičiak, M. (2018). Human capital management in transport enterprises with the acceptance of sustainable development in the Slovak Republic. *Sustainability, 10*(7), 2530. https://doi.org/10.3390/SU10072530

Lee, Y.-K., Chang, C.-T., & Chen, P.-C. (2017). What sells better in green communications: Fear or Hope? *Journal of Advertising Research, 57*(4), 379–396. https://doi.org/10.2501/JAR-2017-048

Leonidou, L. C., Leonidou, C. N., & Kvasova, O. (2010). Antecedents and outcomes of consumer environmentally friendly attitudes and behaviour. *Journal of Marketing Management, 26*(13–14), 1319–1344. https://doi.org/10.1080/0267257X.2010.523710

Lindenberg, S., & Steg, L. (2007). Normative, gain and hedonic goal frames guiding environmental behavior. *Journal of Social issues, 63*(1), 117–137.

Lindenberg, S., & Steg, L. (2013). Goal-framing theory and norm-guided environmental behavior. In *Encouraging sustainable behavior* (pp. 37–54). Psychology Press.

Ling, M., & Xu, L. (2020). Relationships between personal values, micro-contextual factors and residents' pro-environmental behaviors: An explorative study. *Resources, Conservation and Recycling, 156*, 104697. https://doi.org/10.1016/j.resconrec.2020.104697

Liu, P., Han, C., & Teng, M. (2021). The influence of internet use on pro-environmental behaviors: An integrated theoretical framework. *Resources, Conservation and Recycling, 164*, 105162. https://doi.org/10.1016/J.RESCONREC.2020.105162

Mahmood, S., Khwaja, M. G., & Jusoh, A. (2019). Electronic word of mouth on social media websites: Role of social capital theory, self-determination theory, and altruism. *International Journal of Space-Based and Situated Computing, 9*(2), 74. https://doi.org/10.1504/IJSSC.2019.104217

Mehrabian, A., & Russell, J. A. (1974). *An approach to environmental psychology.* The MIT Press.

Meneses, G. D. (2010). Refuting fear in heuristics and in recycling promotion. *Journal of Business Research, 63*(2), 104–110. https://doi.org/10.1016/j.jbusres.2009.02.002

Mikael Klintman, M. B. (2013). Political consumerism and the transition towards a more sustainable food regime: Looking behind and beyond the organic shelf. In *Food practices in transition* (pp. 127–148). Routledge. https://doi.org/10.4324/9780203135921-12

Milfont, T. L. (2007). *Psychology of environmental attitudes.* Unpublished doctoral thesis. The University of Auckland.

Mohd Suki, N., & Mohd Suki, N. (2019). Examination of peer influence as a moderator and predictor in explaining green purchase behaviour in a developing country. *Journal of Cleaner Production, 228*, 833–844. https://doi.org/10.1016/j.jclepro.2019.04.218

Nguyen, H. V., Nguyen, C. H., & Hoang, T. T. B. (2019). Green consumption: Closing the intention-behavior gap. *Sustainable Development.* https://doi.org/10.1002/sd.1875

Prete, M. I., Piper, L., Rizzo, C., Pino, G., Capestro, M., Mileti, A., Pichierri, M., Amatulli, C., Peluso, A. M., & Guido, G. (2017). Determinants of southern Italian households' intention to adopt energy efficiency measures in residential buildings. *Journal of Cleaner Production, 153*, 83–91. https://doi.org/10.1016/j.jclepro.2017.03.157

Rahman, I., Chen, H., & Reynolds, D. (2020). Evidence of green signaling in green hotels. *International Journal of Hospitality Management, 85*, 102444. https://doi.org/10.1016/J.IJHM.2019.102444

Rogers, R. W., & Mewborn, C. R. (1976). Fear appeals and attitude change: Effects of a threat's noxiousness, probability of occurrence, and the efficacy of coping responses. *Journal of Personality and Social Psychology, 34*(1), 54–61. https://doi.org/10.1037/0022-3514.34.1.54

Sangroya, D., Kabra, G., Joshi, Y., & Yadav, M. (2020). Green energy management in India for environmental benchmarking: From concept to practice. *Management of Environmental Quality: An International Journal, 31*(5), 1329–1349. https://doi.org/10.1108/MEQ-11-2019-0237

Schwartz, S. H. (1994). Are there universal aspects in the structure and contents of human values? *Journal of Social Issues, 50*(4), 19–45.

Sharma, K., & Bansal, M. (2013). Environmental consciousness, its antecedents and behavioural outcomes. *Journal of Indian Business Research, 5*(3), 198–214. https://doi.org/10.1108/JIBR-10-2012-0080/FULL/XML

Shaw, D., McMaster, R., & Newholm, T. (2016). Care and Commitment in Ethical Consumption: An Exploration of the 'Attitude–Behaviour Gap'. *Journal of Business Ethics, 136*(2), 251–265. https://doi.org/10.1007/s10551-014-2442-y

Sheth, J. N., Newman, B. I., & Gross, B. L. (1991). Why we buy what we buy: A theory of consumption values. *Journal of Business Research, 22*(2), 159–170.

Shin, Y. H., Im, J., Jung, S. E., & Severt, K. (2018). The theory of planned behavior and the norm activation model approach to consumer behavior regarding organic menus. *International Journal of Hospitality Management, 69*, 21–29. https://doi.org/10.1016/J.IJHM.2017.10.011

Singh, S. K. (2019). Sustainable business and environment management. *Management of Environmental Quality: An International Journal, 30*(1), 2–4. https://doi.org/10.1108/MEQ-01-2019-213/FULL/PDF

Steg, L., Bolderdijk, J. W., Keizer, K., & Perlaviciute, G. (2014). An integrated framework for encouraging pro-environmental behaviour: The role of values, situational factors and goals. *Journal of Environmental Psychology, 38*, 104–115.

Stern, P. C., Dietz, T., Abel, T. D., Guagnano, G., & Kalof, L. (1999). A Value-Belief-Norm Theory of support for social movements: The case of environmentalism. *Human Ecology Review, 6*(2), 81.

Swenson, M. R., & Wells, W. D. (2018). Useful correlates of pro-environmental behavior. *Social Marketing*, 91–109. https://doi.org/10.4324/9781315805795-7

Szabo, S., & Webster, J. (2021). Perceived greenwashing: The effects of green marketing on environmental and product perceptions. *Journal of Business Ethics, 171*(4), 719–739. https://doi.org/10.1007/s10551-020-04461-0

Tang, Y., Chen, S., & Yuan, Z. (2020). The effects of hedonic, gain, and normative motives on sustainable consumption: Multiple mediating evidence from China. *Sustainable Development, 28*(4), 741–750. https://doi.org/10.1002/sd.2024

Tobler, C., Visschers, V. H. M., & Siegrist, M. (2012). Addressing climate change: Determinants of consumers' willingness to act and to support policy measures. *Journal of Environmental Psychology, 32*(3), 197–207. https://doi.org/10.1016/J.JENVP.2012.02.001

Trivedi, R. H., Patel, J. D., & Acharya, N. (2018). Causality analysis of media influence on environmental attitude, intention and behaviors leading to green purchasing. *Journal of Cleaner Production, 196*, 11–22. https://doi.org/10.1016/j.jclepro.2018.06.024

Vining, J., & Ebreo, A. (1992). Predicting recycling behavior from global and specific environmental attitudes and changes in recycling opportunities. *Journal of Applied Social Psychology, 22*(20), 1580–1607. https://doi.org/10.1111/j.1559-1816.1992.tb01758.x

Wang, J., Bao, J., Wang, C., & Wu, L. (2017). The impact of different emotional appeals on the purchase intention for green products: The moderating effects of green involvement and Confucian cultures. *Sustainable cities and society, 34*, 32-42. http://www.elsevier.com/open-access/userlicense/1.0/

Wang, S. T. (2014). Consumer characteristics and social influence factors on green purchasing intentions. *Marketing Intelligence and Planning, 32*(7), 738–753. https://doi.org/10.1108/MIP-12-2012-0146/FULL/XML

Wang, Y., Li, Y., Zhang, J., & Su, X. (2019). How impacting factors affect Chinese green purchasing behavior based on fuzzy cognitive maps. *Journal of Cleaner Production, 240*, 118199. https://doi.org/10.1016/j.jclepro.2019.118199

Wolstenholme, E., Poortinga, W., & Whitmarsh, L. (2020). Two birds, one stone: The effectiveness of health and environmental messages to reduce meat consumption and encourage pro-environmental behavioral spillover. *Frontiers in Psychology, 11*, 2596. https://doi.org/10.3389/FPSYG.2020.577111/BIBTEX

Yang, X., Chen, S. C., & Zhang, L. (2020). Promoting sustainable development: A research on residents' green purchasing behavior from a perspective of the goal-framing theory. *Sustainable Development, 28*(5), 1208–1219. https://doi.org/10.1002/sd.2070

Yang, X., & Zhang, L. (2020). Diagnose barriers to sustainable development: A study on "desensitization" in urban residents' green purchasing behavior. *Sustainable Development, 28*(1), 143–154. https://doi.org/10.1002/SD.1978

Yarimoglu, E., & Binboga, G. (2019). Understanding sustainable consumption in an emerging country: The antecedents and consequences of the ecologically conscious consumer behavior model. *Business Strategy and the Environment, 28*(4), 642–651. https://doi.org/10.1002/BSE. 2270

Zabkar, V., & Hosta, M. (2013). Willingness to act and environmentally conscious consumer behaviour: Can prosocial status perceptions help overcome the gap? *International Journal of Consumer Studies, 37*(3), 257–264. https://doi.org/10.1111/J.1470-6431.2012.01134.X

Zelezny, L. C., & Schultz, P. W. (2000). Psychology of promoting environmentalism: Promoting environmentalism. *Journal of Social Issues, 56*(3), 365–371. https://doi.org/10.1111/ 0022-4537.00172

Zhang, L., Li, D., Cao, C., & Huang, S. (2018). The influence of greenwashing perception on green purchasing intentions: The mediating role of green word-of-mouth and moderating role of green concern. *Journal of Cleaner Production, 187*, 740–750. https://doi.org/10.1016/J.JCLEPRO. 2018.03.201

Zheng, S., Wang, J., Sun, C., Zhang, X., & Kahn, M. E. (2019). Air pollution lowers Chinese urbanites' expressed happiness on social media. *Nature Human Behaviour, 3*(3), 237–243. https://doi.org/10.1038/s41562-018-0521-2

Business Analytics and Sustainability

Safwat Al Tal

1 Introduction

The information era, which began toward the close of the twentieth century and continued into the twenty-first, made information freely available to academics, scholars, and corporations. The desire for increased transparency from many stakeholders has fueled this increase in information accessibility, particularly in the corporate sphere. Around the same time, the concept of sustainability evolved, a company strategy aimed at maximizing long-term shareholder value by seizing opportunities and minimizing economic, environmental, and social risks. The confluence of these two sectors paved the way for big data to transform how businesses with environmental impacts can take meaningful action toward sustainability.

2 What Is Business Analytics?

Let us start with what is Analytics? Analytics is the systematic computational analysis of data or statistics. This analysis when optimized into meaningful business-related insights that can help it grow and improve, then it becomes Business Analytics. The emergence of big data has increased organizations' demand for business analytics, defined as the "extensive use of data, statistical and quantitative analysis, explanatory and predictive models, and fact-based management to drive decisions and actions (Davenport & Harris, 2007). However, in reality it is never quite this simple! In analytics, it is often more challenging; as data requires plenty of

S. Al Tal (✉)
Department of Business Analytics, Faculty of Business, Abu Dhabi, UAE
e-mail: saltal@hct.ac.ae

cleansing before it can even prove to be useful! In addition, on many occasions and after doing different levels of analysis we discover that not all output is helpful! Some of it could be just noise!

There is a widely spread misconception about analytics in the minds of many people in which it is thought of as a glamorous way of showing beautiful charts graphs and dashboards! However, these visual aspects are an essential part of it but analytics is a lot more than that! To explore what business analytics is about, think of the following examples of scenarios that would help you understand the scope of business analytics.

Let us say you work at a credit card company and you want to analyze data about your customers to determine who might subscribe to a credit card offer that would allow you to spend more time and energy on targeting those specific customers. You will be analyzing data about past transactions, process them into a present perception of credit card customers so that you can predict their future behavior of them (who might subscribe to a credit card offer).

Let us think of another scenario, imagine that you are working at a tech company and you want to review data about your employees to understand why they leave and then take action to minimize employees turnover! The same scheme of analyzing data related to the past "what has happened with the employees who left? Processing it into a present perception that will tell you "why it happened" so you can predict an outcome in the future "who might leave in the future?" As you can see, these business scenarios reveal how you could analyze data in different business scenarios to turn out or churn meaningful insights by analyzing the past to understand the present so you can predict the future; and this is the essence of Business Analytics. It focuses on developing new insights and understanding of business performance based on data and statistical methods (Oliver et al., 2018).

When we talk about Business Analytics there are different terms that you would hear around analytics, terms such as business analysis, business intelligence, data analysis, data analytics, and data science are often confused and used interchangeably. While there is a lot of overlap in the concepts, there are subtle differences between them but for the most part they all support the same goal of turning data into useful insights.

3   Types of Business Analytics

We tend to use the word analytics frequently but in reality, we have a number of different flavors of analytics to work with and each of these has its own specific purpose. Analytics is broadly classified into four categories, those are:

1. Descriptive Analytics: Typically look into the past or the present, have questions along the lines of tell me what happened and why it happened as well as tell me what is happening right now and then why it is happening.

2. Predictive Analytics: Ask questions such as tell me what is likely to happen and why.
3. Discovery Analytics: These are questions along the lines of looking for something important, asking very specific questions, in other words, we tend to be mining through our data and looking for interesting patterns and correlations.
4. Prescriptive Analytics: Tell us not only what is going on or what might have happened or what is likely to happen but what we should do about it and other things related to specific business actions we should be taking.

We take the outputs of all of these categories of analytics and we feed them into our prescriptive analytics engines to find all the insights of our business. This entire analytics process is essential to us, the whole four categories need to work together in order to give the greatest business benefit that we are after (Pochiraju and Seshadri, 2019).

Now let us talk about the analytics lifecycle and you can think about this like the scientific method, for analytics it is a tried and true way of doing things and the most common lifecycle is what is called the CRISP-DM, which stands for Cross-Industry Standard Process for Data Mining, this is an industry-proven way to guide your data mining efforts. As a methodology, it includes descriptions of the typical phases of a project, the tasks involved with each phase, and an explanation of the relationships between these tasks. The CRISP-DM starts with understanding your business problem to be followed by data understanding and getting the data ready, after that the modeling starts to be followed by evaluating the model and deploying it in the real world.

Let us talk about these five phases of the life cycle in a little more detail:

1. Business understanding, the main part of this phase is to start with business problems, we need to have business goals in mind, some examples might be to optimize pricing to boost revenue or to segment customers to tailor product offers to them or pinpoint bottlenecks and failure points in our supply chain.
2. Data understanding and a lot of what we are doing here are looking at what data we have and what data we need and trying to cover some of those gaps and then when we get the data we might ask questions related to it:

   • Availability
   • Quality
   • Granularity: How deep or detailed does it go?
   • Frequency: How often does it get updated and so on?

   Now as we are trying to understand and explore data oftentimes we use a sandbox which is a safe space to explore data so we do not mess up with what is called "production" and that is where all the live data is.

3. Data preparation is the next phase and oftentimes this can be the most time and effort-consuming phase as we need to cleanse and scrub the data to get ready for further modeling and analysis, so here is an example:

Look at the following data in the table, what do you notice?

Name	Gender	City	State	DOB
Michael	Male	Utica	New York	02–12-88
Rose	Female		NY	05 April 1977

To get some data like this is no surprise!! If you see something like where you have two customers one of them has a city that is missing, the states are in different formats, and also the date of birth!! So what you have to do is to perform what we call data cleansing, also referred to as data cleaning or data scrubbing, which is a requirement to get this data in a good place in order to do the next step to work on your model. In addition, you might need to go back to the technology teams and say let's make sure we can restrict the inputs that we're getting so we can do the analysis with good quality data. After data preparation we're going to go to the:

4. Modeling phase, but first what actually is a model? A very common definition of a model is a simplified description of a system or process to assist calculations and predictions. To simplify this think of a model as something that mimics the real world! Think of the Lego model (plastic construction toys) you used to build when you were a little child and remember how they looked like the White House or the pyramids of Giza in Egypt! That was a model of the real thing and as we try to make predictions we try to build a model and we try to see if we can get not only a good model but also a simple one to help us make our predictions, here is an example of a model; a model that predicts the likelihood that a car insurance customer will get into an accident in the next year, here we have a model that makes that prediction to mimic what might happen in the real world. So in this modeling phase a few things we are going to do are:

   – Exploratory analysis on the data.
   – Variable selection to figure out what variables should be included in our model.
   – Model selection and the fine tuning of it.

   As for the modeling tools, one of the most common tools that we can use in the modeling phase are "Python" and "R". In many ways, the two open source languages are very similar. Free to download for everyone, both languages are well suited for data science tasks—from data manipulation and automation to business analysis and big data exploration. The main difference is that Python is a general-purpose programming language, while R has its roots in statistical analysis. Increasingly, the question isn't which to choose, but how to make the best use of both programming languages for your specific use cases. After defining our business question and then understanding, preparing the data and building our model then we go to the.

5. Evaluation and deployment phase, where we will ask questions like how effective is the model? Is it working well? Are the predictions fairly accurate? And if so, are we prepared to launch it?

4   Popular Analytics Tools

Data analysis is a core practice of modern business analytics. Choosing the right analytics tool is challenging, as no tool fits every need a few of the most popular ones that we might use in business analytics are:

– Microsoft Excel to help us explore and analyze smaller datasets.
– Tableau desktop to help us visualize our data using dashboards.
– Python programming language to help us build these models to make predictions that we just talked about and then we could use.
– SQL to allow us to communicate and interact with databases.

So it is not uncommon for job postings in business analytics to cover a lot of these tools but we can mention a couple of other alternative tools for visualization and for model building that are not listed above, for example in place of Tableau desktop we can use Microsoft power BI and instead of Python, we might use the "R" programming language.

5   Sustainability in Business

In general, sustainability describes meeting our own needs without compromising the ability of future generations to meet their own needs (WCED, 1987). In addition to natural resources, we also need social and economic resources. Business-wise, sustainability is a business approach to creating long-term value by taking into consideration how a given organization operates in the ecological, social, and economic environment Gao et al. (2020). What could be an example of sustainability in business is Google's high electricity bills and its impact on the environment, which has made Google interested in renewable energy for a long time. Renewable energy projects at Google range from a large solar power installation on its campus, to the promotion of plug-in hybrids, to investments in renewable energy start-up companies like e-Solar. Also, Google has worked on making its data centers more energy efficient, meaning that they consume less electricity while still handling the same amount of data requests. The definition of sustainability in business can be recognized by what was written by Rick Needham, the green business operations manager at Google, on the company's blog: "We're aiming to accelerate the deployment of renewable energy—in a way that makes good business sense, too." Accordingly, we can define a sustainable business as one that operates in the interest of all current and future stakeholders in a manner that ensures the long-term health and survival of the business and its associated economic, social, and environmental systems.

The question yet to be asked is, to which extent businesses are integrating sustainability into their corporate DNAs? Is there reliable evidence that quantifies the value of a company integrating sustainable development into its corporate DNA? It turns out that the answer is, "Yes."

For that yes, we need to look into the work of business sustainability guru Bob Willard who has worked for 34 years with IBM, Bob completed first a master's and then a doctorate on the business case for sustainability at the University of Toronto. He is the author of several books and is considered a global expert on this topic. Now that we know what we are talking about, how do we know whether integrating sustainability into the corporate's DNA will actually help steward the five capitals in ways that generate revenue and mitigate risk? After years of research and hundreds of case studies, Bob Willard's research shows that if a typical company were to use best-practice sustainability approaches already being used by real companies, a Small or Medium Enterprise (~6 employees and $one million dollar revenue) could improve its profit by at least 51% and a large manufacturing/distribution company (of several thousand employees and a $500 million revenue) could improve its profit by at least 81% compared with if it did nothing. There are 7 bottom line benefits that align with current evidence about the most significant sustainability-related contributions to profit. Increased revenue: 9%, Reduced energy: 75%, Reduced waste: 20%, Reduced materials: 10%, Increased productivity: 2%, Reduced turnover (attrition): 25%, Avoid a potential 16% to 36% erosion of profits due to reduced risks (revenues and expenses). It is true! Just by doing what others are already doing a business can increase its profit by 51–81% within 3–5 years. And, keep in mind this is only about best practices; innovation and creating better products and services can take you even further. Sustainability is just smart business. The numbers we've discussed are just the tip of the iceberg when it comes to the business case for sustainability (Willard, 2002).

5.1 What is Corporate Sustainability?

The majority of individuals are still perplexed by the concept of business sustainability. There has been a tidal wave of firms committing to "sustainability" in the previous 2 years. These firms may set net-zero carbon objectives, diversify their workforce, or enter new, more environmentally friendly lines of business, and this is only the beginning of the wave.

As corporations face pressure from social movements and environmental challenges, interest in sustainability is projected to rise even more over the coming decade. "So, what do you mean by sustainability?" people frequently inquire.

To contribute to sustainable development, businesses should create wealth to reduce poverty, but do so without harming the natural environment. In this way, businesses help our world today and ensure that future generations can also thrive, in practice, this means that businesses must consider three key things in their operations (Bansal and Agarwal, 2021; Kaur, 2019):

1. Human rights and social justice

   Sustainability requires businesses to recognize their impact on the people they employ and the surrounding communities. This recognition means committing to fair wages, just and ethical treatment, and clean safe environments.

   For example, the clothing retail industry racked up billions in unpaid bills during COVID-19 because of plummeting prices and unsold garments, leaving millions of garment workers in desperate conditions. Yet, some brands continued to pay suppliers, even though their own incomes had declined steeply.

2. Natural resource extraction and waste

   Businesses often rely on natural resources such as land, water, and energy. While many natural resources can renew or "regenerate," in cycles businesses need to respect these cycles, by using natural resources at the speed at which they regenerate. For example, companies can reduce their resource extraction by using recycled or repurposed products and make their operations more efficient by reducing waste. By doing so, they contribute to the "circular economy" wich will lead to quick gains, i.e. a vegetable processing plant saved $300,000 by simply capturing beans falling off a processing line.

3. Short- and long-term thinking

   Businesses face intense pressure for immediate profits, but sustainability requires investing in technologies and people for the future, even though financial benefits show up much later. Companies are used to longer-term thinking for capital investments, but a sustainability orientation applies this logic to investments in people and society. For example, some fossil fuel companies have reimagined themselves as energy companies, even though major investments in renewable energies are less profitable in the short run than their oil, gas, or coal operations. They recognize that climate change requires them to build new capabilities and sources of energy.

6   How Does Corporate Sustainability Differ from Corporate Social Responsibility?

Many terms exist to describe companies' social and environmental initiatives. Corporate Social Responsibility (CSR) is the most common; others include environmental, social, and governance (ESG), shared value, the triple bottom line, and managing environmental impacts (Soyer and Asan, 2022).

Sustainability is considered as the most complete and powerful of these related concepts. That is because sustainability asks managers to take a "system's view." A system's outlook recognizes that companies are part of a larger social and environmental system, systems change, and today's actions must consider the future.

CSR emphasizes a company's ethical responsibilities. However, what is ethical for one person or company may not be seen as ethical by another. For example, some people see a minimum wage as being responsible, whereas others see a higher "living wage" as an ethical choice. Corporate sustainability emphasizes science-based principles for corporate action. A corporate sustainability lens would set a wage in which people could meet their basic needs, which will vary from place to place. Additionally, CSR generally does not speak to fairness across generations; it focuses more on the present. However, do not get too lost in the definitions; ultimately, all of these terms ask businesses to think about the broader world in which they operate, and not just on short-term self-interest (Correia, 2019; Soyer and Asan, 2022).

7   Why Is Corporate Sustainability Important?

Business is a powerful actor in society, with some businesses being larger than some governments. For example, Amazon's revenues in 2019 were $US281bn: larger than Pakistan's GDP. Businesses now have so much power that executives can choose to create a better life for all or just a few.

Society is also pushing companies to invest in sustainability. Many governments, citizens, and other stakeholders want to see companies showing concern for their communities. Failing to do so can mean losing the social license to operate, which is society's trust in a company.

Additionally, companies can benefit in the long term from being green and good. Evidence shows that financial benefits come in many forms. For example, reducing waste, through energy efficiency investments, often produces savings (Gao et al., 2020; Gittel, 2012)

Investors increasingly look for companies that have higher Environmental, Social, and Governance "ESG" ratings, as a way of managing risks.

Creative and committed individuals seek out employers committed to sustainability and are even willing to take a lower salary if such a commitment is sincere.

But, let us face it, let us be honest. It us not just about making money when it comes to sustainability. It is also a vision of what successful corporate leaders wish to see in the world they build. They envision a society in which everyone may thrive while living on a durable and biodiverse earth. They do not want to live in a world where only a few people live happily while others suffer from disease and waste (Wen, 2014; Ray, 2017).

8   The Triple Bottom Line TBL/3BL

The triple bottom line (TBL or 3BL) is a concept that is used a lot when speaking about sustainable development, and particularly sustainability in business. John Elkington, a global authority on corporate responsibility and sustainability coined the phrase in a book in 1997. His argument was that the methods by which companies measure value should include not only a financial bottom line (profit or

Fig. 1 the triple bottom line
concept (Elkington, 1997)

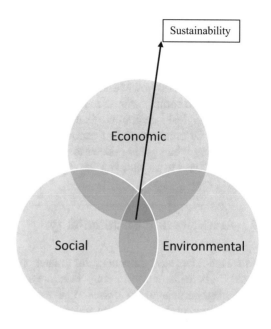

loss), but a social and environmental one as well (Elkington, 1997). Sustainability is typically defined as the place where economy, social realities, and environmental health overlap as shown in the following Fig. 1. The concept of the triple bottom line mainstreamed the idea of sustainability as including people, planet AND profit. It helped business to understand that long-term sustainability of an organization required more than just financial equity. It also helped to clarify that when businesses were considering what sustainability meant for them, it did not mean they had to give up the notion of financial success (Willard, 2002).

But this overlapping circles image of the triple bottom line can convey a lot more. The circles are all the same size. Does this indicate that the economy is the same relative size, or value, as the other two circles, which deal with society and the environment? Can we trade say "2 social and 3 environment for 5 economy" as long as we stay in the overlapping bit in the middle (sustainability)? Science tells us that, left to its own devices, the planet operates in a balanced way. We call this the cycles of nature and they are powered by energy from the sun. Science also tells us that matter is not created or destroyed, while laws of thermodynamics tell us that everything tends toward dispersal (principle of entropy). Because plant cells are, for all intents and purposes, the only cells that can produce a structure from energy, photosynthesis is the process by which matter is structured on our planet. This is why we say that photosynthesis pays the bills. Without it, creation of structure from energy would not occur, and entropy could rule the day. How does this help us understand the triple bottom line? Plant cells belong to the environment circle of the triple bottom line. If plant cells are the original creators of structure, then this is the circle on which everything else depends, or in which everything is embedded.

Everything comes from nature at some point. Society, which is related to the social circle of the triple bottom line, exists within the environment and the economy is a by-product of society (Barreto, 2010).

Instead of three overlapping circles, we have three nested circles, where the economy is a wholly-owned subsidiary of the environment. To achieve sustainability, we need to comply with social and environmental conditions: meet human needs within ecological constraints. Does this mean that business has to put financial gain last? Of course not! But economic decisions are part of a strategy to make more money while getting closer to social and ecological sustainability. The economy is a means to an end. Not the end itself.

9 Analytics and Sustainability

As stakeholders have demanded for greater openness and organizations have recognized the benefits of considering corporate sustainability, sustainability has become increasingly crucial to businesses. As a result, there has been a significant increase in transparency, both in corporate statements and in annual reports, in which corporations outline their environmental initiatives. These documents and reports are of great value because they include a plethora of information that may be investigated and analyzed.

Previously, the contents of these reports were examined using manual methods; however, there is significant potential for automated analysis through Business Analytics methods and tools.

The opportunities associated with data and analysis in different organizations have helped generate significant interest in Business Analytics, including all the techniques, technologies, systems, practices, methodologies, and applications that analyze critical business data to help an enterprise better understand its business and market and make timely business decisions according to what we have explained previously, Business Analytics includes business-centric practices and methodologies that can be applied to various high-impact applications in different sectors (Chen & Storey., 2012).

Companies want to take advantage of the business benefit of sustainability now that they have discovered it and sustainability analytics assists them in achieving this goal. Sustainability analytics may help businesses evaluate the cost, effect, and performance of their past and current sustainability activities, as well as forecast future conditions and requirements, allowing them to uncover hidden value and develop a more resilient business. Companies in order to gain their deep insights they need to guide their sustainability-related initiatives and improve their overall resource efficiency by collecting and analyzing data on a wide range of sustainability-related factors, for example, energy and resource use, greenhouse gas emissions, and supply chain performance, and much more within different sectors (Gittel, 2012).

Advances in computing and data science now allow companies to conduct real-time (or near real-time) sustainability analysis on vast quantities of data within the three dimensions of time by:

1. Analyzing the past
   Even in today's real-time world, analyzing past performance can produce valuable insights. Examine the historical performance metrics you are currently measuring and ask whether they are still relevant to your sustainability goals.
2. Understanding the present
   Tools and techniques for analyzing sustainability have advanced by leaps and bounds. Can your current systems generate insights in real time (or near real time)? Are they fully automated and linked to performance dashboards? Do you have clear mechanisms for governing and managing sustainability data? If you answered "no" to any of these questions, it might be time for an upgrade.
3. Modeling the future sustainability landscapes
   The latest tools also include modeling and scenario analysis that can help you understand the complex interplay of economic, social, and environmental factors that affect your future sustainability strategies and performance.

10   Conclusion

At the end of the twentieth century and into the twenty-first century, the information age has allowed for information to be widely accessible to researchers, scholars, and businesses.

This increase in accessibility of information has especially been spurred on in the corporate world by a desire for greater transparency from various stakeholders. Around the same time emerged the concept of sustainability, a business approach that aims to create long-term shareholder value by embracing opportunities and managing economic, environmental, and social risks. The intersection of these two fields set the stage for the need to maintain and publish an increased amount of electronic information. Much of the information is made available through governmental regulations that require corporations to provide information about their operations. For example, through annual reports, companies will often share information about how the company performed relative to expectations. There has also been an increase in voluntary disclosure through corporate sustainability reports, which allow companies to further describe their environmental sustainability efforts. The disclosure of business information is impacted as topics such as sustainability and community empowerment have become more important in the eyes of stakeholders. While companies vary on the details of the information that is published, many companies will provide a view of the overall impact. Therefore, these reports contain a wealth of information that can be studied and analyzed. While there are many ways to analyze these reports, it is important to explore the potential of computational analysis because of the sheer amount of information that is available. Computational methods allow for much quicker analysis and can remove the potential bias that is present in human analysis. It is also essential to note that using programs to analyze corporate information has become more popular with the increased computing power of machines. The output of these programs can further be used to analyze information in more detail. It also allows for flexibility in analysis, as any disclosure can be analyzed, not just annual reports.

References

Bansal, T., & Agarwal, D. (2021). *Network for business sustainability.*

Barreto, I. (2010). Dynamic capabilities: A review of past research and an agenda for the future. *Journal of Management, 36*(1), 256–280.

Chen, C., & Storey. (2012). Business intelligence and analytics: From big data to big impact. *MIS Quarterly, 36*(4), 1165. https://doi.org/10.2307/41703503

Correia, M. S. (2019). Sustainability: An overview of the triple bottom line and sustainability implementation. *International Journal of Strategic Engineering (IJoSE), 2*(1), 29–38. https://doi.org/10.4018/IJoSE.2019010103

Davenport, T. H., & Harris, J. G. (2007). *Competing on analytics: The new science of winning.* Harvard Business Press.

Elkington, J. (1997). *Cannibals with forks – Triple bottom line of 21st century business.* New Society Publishers.

Gao, Y., Tsai, S., Du, X., & Xin, C. (Eds.). (2020). *Sustainability in the entrepreneurial ecosystem: Operating mechanisms and Enterprise growth.* IGI Global. https://doi.org/10.4018/978-1-7998-3495-3

Gittell, R. (2012). *The sustainable business case book.* University of New Hampshire/University of New Hampshire Whittemore School of Business/Saylor Foundation. ISBN 13: 9781453346778.

Kaur, T. (2019). Corporate sustainability: The base of corporate social responsibility – A case study of TCS. In I. Management Association (Ed.), *Corporate social responsibility: Concepts, methodologies, tools, and applications* (pp. 955–971). IGI Global. https://doi.org/10.4018/978-1-5225-6192-7.ch048

Oliver, M., Fay, M., & Vom Brocke, J. (2018). The effect of big data and analytics on firm performance: An econometric analysis considering industry characteristics. *Journal of Management Information Systems, 35*(2), 488–509.

Pochiraju, B., & Seshadri, S. (2019). *Essentials of business analytics an introduction to the methodology and its Applications: An introduction to the Methodology and its Applications.* Springer. https://doi.org/10.1007/978-3-319-68837-4

Ray, N. (Ed.). (2017). *Business infrastructure for sustainability in developing economies.* IGI Global. https://doi.org/10.4018/978-1-5225-2041-2

Soyer, A., & Asan, U. (2022). *Analyzing the relationship between corporate governance, CSR, and sustainability.* IGI Global. https://doi.org/10.4018/978-1-7998-4234-7

Wen, J. (2014). A business analytics approach to corporate sustainability analysis. In *Master of environmental studies capstone projects.* p. 62.

Willard, B. (2002). *The sustainability advantage: Seven business case benefits of a triple bottom line.* New Society Publishers.

World Commission on Environment and Development. (1987). *Our common future.* Oxford University Press.

Data Science and External Audit

Ahmad Faisal Hayek

1 Aspects of Using Big Data Analytics in External Audit

1. Assess fraud risk.
2. Audit evidence.

1.1 Assess Fraud Risk

BDA can help assess fraud risks and perform specific audit procedures to address those risks. For example, in identifying specific journal entries with unique risk profile. This identification results in the application of further audit procedures and analysis to support auditors' assessment of fraud risks.

Some benefits of using data analysis in fraud detection are (Rajvanshi, 2016):

- Identify low incidence of events: Analytics can be used to find fraud cases that are not very obvious and are low incidence, and then use predictive analytics on these fraud cases to analyze them further.
- Enterprise-wide solution: Analytics takes an enterprise-wide global perspective, which helps in detecting fraud by associating related information in a company.
- Data integration: A very useful way of detecting fraud in a company is getting as much data as possible. Analytics makes it possible by gathering data from different sources along with some outside data that may have some prediction capability.

A. F. Hayek (✉)
Department of Accounting, Faculty of Business, Higher Collogues of Technologies, Abu Dhabi, United Arab Emirates
e-mail: ahayek@hct.ac.ae

- Utilizing unstructured data: Sometimes the unstructured data is not properly stored and it contains the most valuable information related to fraud. Analytics help store this data and bring it to some value.

Artificial Intelligence and machine learning techniques for fraud detection can be divided into Supervised and Unsupervised learning. These methods predict what sections of customers or users are more likely to commit fraud by giving out a score or rule (Balios et al., 2020).

Supervised Learning It is a type of machine learning algorithm which is used to detect fraud cases that follow a pattern already been identified by the system. In this method, a sample from the entire data is taken and the fraudulent cases are separated from the rest. By training the model to identify the fraudulent cases, these models are then used to identify new fraud cases by applying them to new data.

Unsupervised Learning It is used to detect any new type of fraud that may occur which was never seen before. This, unlike supervised learning does not use labeled records.

1.2   Audit Evidence

As per International Standards on Auditing, standard 500 "Audit evidence," the auditor should obtain sufficient appropriate audit evidence to be able to draw reasonable conclusions on which to base the audit opinion.

Appropriate means reliable and relevant, while sufficient is related to the volume and variety, and Big Data can contribute to this aspect. When traditional data are not adequately "reliable" or "relevant," then more evidence from Big Data can be useful. Furthermore, Big Data can be more reliable and relevant than traditional sources, even though "noise" may impact on their reliability.

As far as reliability is concerned, some types of Big Data can contribute to evaluating the reliability of traditional audit information. Besides, Big Data from external sources can provide crucial nonfinancial evidence which can be used to assess financial accounts. For example, when a product receives adverse comments on Social Media, while the Sales registered in the financial statements have increased, then this could be a sign for further investigation.

Unlike traditional audit practices, the technology-enabled audit comes with a higher quality of audit evidence, which is derived from many new sources, including big data, exogenous data, the ability to analytically link different processes, database-to-database confirmation, and continuous monitoring alerts.

The amount of resources used could be reduced if Big Data Analytics has the potential to replace existing labor-intensive parts of the financial statement audit. For example, testing the value of the inventory from retailer is generally done by obtaining evidence examination which is labor intensive. Brown Liburd and

Vasarhelyi stated that using the information of radio frequency identification (RDFI) chips for validating inventory could make the process more labor efficient.

Big Data supported analytical tools do not just examine the entire population of available data, but also incorporate other unstructured data to establish a relationship between every data examined and draw useful insights from them.

For example, data analytic tools can analyze entire financial data at different dimensions like date, time, purpose, transaction types, transaction value, business type, customer type, geography, standards-based, and so on.

The analytic procedure is a significant part of the audit process, which includes analyzing data to find out any plausible relationship between financial and nonfinancial data. At the same time, BDA is the use of data, information technology, statistical analysis, quantitative methods, and mathematical or computer-based models to help managers gain improved insight about their operations, and make better, fact-based decisions (Jesus, 2018).

Analytical procedures, according to Auditing Standards 2305 (PCAOB, AS 2305 2016), are an important part of the audit process and mainly consist of an analysis of financial information made by a study of believable or plausible relationships among both financial and nonfinancial data. These analytical procedures could be as basic as scanning (viewing the data for abnormal events or items for further examination) to more complex approaches (not clarified by the standards, except that the approach should enable the auditor to appropriately develop an expectation and subsequently examine these expectations to the reported results).

Big Data Analytics (BDA) that is utilized by client management and their accountants has been defined as "the use of data, information technology, statistical analysis, quantitative methods, and mathematical or computer-based models to help managers gain improved insight about their operations, and make better, fact-based decisions."

The Auditing standards define the task for analytical procedures in each of the three audit phases (risk assessment/planning phase, substantive testing, and review phase), but are noncommittal about which techniques auditors should undertake to achieve these objectives. Hence, whether an auditor employs more complex analytics such as belief Functions or "traditional analytical procedure" techniques such as ratio analysis would seem to depend on the auditor's own knowledge and less so on the standards. It has also been proposed that any adoption by the external audit profession of either advanced analytics or big data would be due to market or business forces exogenous to the firms (Raphael, 2017).

2   Audit Data Analytics

Audit Data Analytics (ADA): is the science and art of using analysis, modeling and visualization to discover and analyze patterns, anomalies, and other information for the purpose of planning and performing an audit.

Audit Data Analytics (ADAs) help auditors discover and analyze patterns, identify anomalies and extract other useful information from audit data through analysis, modeling, and visualization. Auditors can use ADAs to perform a variety of procedures to gather audit evidence, to help with the extraction of data and facilitate the use of audit data analytics, and as a tool to help illustrate where audit data analytics can be used in a typical audit program.

3   Application of ADA

- Identifying and analyzing anomalies in the data.
- Identifying and analyzing patterns in the data including outliers.
- Building statistical (e.g., regression) or other models that explain the data in relation to other factors and identify significant fluctuations from the model.
- Synthesizing pieces of information from disparate analyses and data sources into wholes that are greater than the sum of their parts for purposes of the overall evaluation.

ADA mode can be exploratory or confirmatory:

Exploratory mode	
When	Planning
Question	What is going on here? Does the data suggest something might have gone wrong? Where do the risks appear to be? What assertions should we focus on?
Approach style	Bottom-up, inductive, few starting assumptions, assertion-free
Methods	Graphical visualizations used to discover patterns in and understand the data— Possibly several to get different viewpoints
Results	Identified risks, areas of focus, potential models for confirmatory stage
ADA examples	– Cluster analysis – Test and data mining – Scatterplots matrices – Line chart – Spread chart – Needle graphs – Small multiples of graphics – Heart maps – Treemaps – Relationship maps

Confirmatory mode	
When	Performance
Question	Does the data conform with and thus confirm my model for what ought to be?
Approach style	Top-down, deductive, model-driven, starts with the development of model based on assertions to be tested
Methods	Comparison of actual data to model taking into account materiality, desired assurance, and assertions being tested; more mathematical than graphical
Results	Identified anomalies, unexpected patterns, outliers, and other significant deviations
ADA examples	– Analytical procedures (regression analysis, ration analysis, reasonableness test) – Recalculations – Journal entry testing – Traditional file interrogation

4 Audit Data Analytics Techniques

- Ratio Analysis: use to calculate ratios using financial and nonfinancial data to gain a high-level understanding of entity operations.
- Sorting: use the software to sort financial and nonfinancial data by categories to identify outliers.
- Trend Analysis: use software to evaluate changes and trends in financial and nonfinancial data over time.
- Matching: use software to electronically match items from various sources on a predetermine characteristics to identify errors and unexpected differences.
- Comparison: use year-over-year comparisons in accounts balances and other nonfinancial data.
- Forecasting: use software to extrapolate past patterns onto future period.
- Predictive Analysis: use software to apply predicated patterns to existing data. Enables auditors to identify situations that have deviated from expectations.
- Cluster: use software to group data into natural clusters. Enables auditors to identify outliers or observations with unique risk characteristics.
- Regression Analysis: use software to understand existing relationships between various characteristics. Enables external auditors to make expectations as to current period balances.
- Process Mining: use software to identify deviations from expected process flow and patterns.

5   Audit Data Analytics Framework

Framework is defined by three Business Analytics dimensions; Domain, Orientation, and Techniques:

(a) Domain: The domain is the environment where audit teams apply analytics like client' enterprise and management.

- Pre-engagement Activities
- Planning
- Compliance Testing
- Substantive Testing and Review
- Opinion Formulation and Reporting
- Continuous Activities.

(b) Orientation: The orientation refers to the vision of analytics descriptive, predictive, or prescriptive.

(c) Technique: The technique is the analytical approach or method (Appelbaum et al., 2015).

- Qualitative or Quantitative
- Deterministic or statistical
- Based on unstructured, semi-structured, or structured data

In addition to the above frameworks, ADA mode can be Expectation, Structural and Multivariate techniques:

- Expectation techniques: An empirical relationship among several accounting numbers or some other important quantitative measures of business operations and is inferred from the archive of historical records.
- Structural techniques look for various structural properties in the historical records. A very popular recent example is process mining.
- Multivariate techniques: The primary objective of multivariate techniques is to develop relationships between or among variables/features under study.

6   Performing Audit Data Analytics (ADAs)

6.1   Plan the ADA

The auditors consider the following when planning the ADA:

- Determine the financial statement items, transactions, accounts, or disclosures, and related assertions and the nature, timing, and extent of the population to which the ADA will be applied.

- Determine whether ADA is to be used in performing a risk assessment procedure, a test of controls, a substantive analytical procedure, a test of details, or in procedures to help form an overall conclusion from the audit.
- Select the techniques, tools, graphics, and tables to be used. Ration Analysis, sorting, trend analysis, matching, comparison, forecasting, predict analysis, cluster analysis, Regression analysis, and process mining.

6.2   Access and Prepare the Data for Purposes of the ADA

This step concentrates on obtaining the data from the entity's ERP or another data source and preparing the data for analysis.

6.3   Consider the Relevance and Reliability of the Data

Data's relevance (the extent to which it relates to the purpose of the ADA) and reliability (the extent to which the data is accurate, complete, and precise) is affected by the data's nature, source, format, timing, extent, and level of aggregation.

6.4   Perform the ADA

How the ADA will be performed depends both on the technique used (for example, regression analysis or trend analysis) and on the purpose of the ADA (for example, preliminary analytical procedure vs. substantive test).

Evaluate the results and conclude on whether the purpose and specific objectives of performing the ADA have been achieved.

7   BDA's Tools and Techniques

The more forward looking the task and the more varied and voluminous the data (big data), the more likely the analysis will be prescriptive or at the very least, predictive. Advanced or more complex BDA may be defined as "Any solution that supports the identification of meaningful patterns and correlations among variables in complex, structured and unstructured, historical, and potential future data sets for the purposes of predicting future events and assessing the attractiveness of various courses of action.

Advanced analytics typically incorporate such functionality as data mining, descriptive modeling, econometrics, forecasting, operations research, optimization, predictive modeling, simulation, statistics, and text analysis".

1. Artificial intelligence
2. Data analytics
3. Machine learning application
4. Data visualization
5. Data mining

7.1 Artificial Intelligence

The original goal with the creation of AI was to make computers more capable of independent thinking. AI uses machines that can interpret and learn from external data. Artificial intelligence is a "computing system that exhibits some form of human intelligence, which covers several interlinked technologies, including data mining, machine learning, speech recognition, image recognition, and sentiment analysis."

Integrating AI in each step of auditing process will remove the repetitive tasks common in the process and make analyzing large volumes of data to have an in-depth understanding of the business operation easier for auditors. Making it easier to concentrate on activities that will bring utmost value to the clients. As assessing the risk of material misstatement is a crucial part of auditing. Auditors are expected to carry out tests on the transactions to make certain that there are no misstatements, for if financial impacts are not accurately recorded, financial statements are bound to be materially misstated. If unauthorized transactions and/or other irregularities are not detected in time, it may be challenging for auditors to capture such later. AI-based tools in auditing make detecting such high-risk transactions easy. This which manual auditing may sometimes not capture fully as a result of sample population testing, unlike the AI technology that allows for full population testing.

Patterns of Artificial Intelligence
Hyper Personalization This pattern uses machine learning to develop a profile unique to each individual. This profile will adapt over time and is used to provide personalized content unique to each user instead of grouping users into categories.
Autonomous System An autonomous system can complete a task, goal, and interact with its surroundings with minimal or no human involvement. This autonomous system can have both hardware and software components, but the overall goal is to minimize human labor. Autonomous systems are used in cars, airplanes, boats, and more which provide information with minimal human involvement.
Predictive Analytics and Decision Support This process involves using cognitive approaches and machine learning to determine patterns that can help predict future outcomes. Predictive analytics is used in projection methods such as forecasting to help humans make better decisions.
Conversation and Human Interaction The objective of conversation and human interaction is to enable machines to interact with humans the way that humans interact with each other. The ability of machines to communicate with humans includes voice assistants, chatbots, and the generation of text, images, and audio.

Anomaly and Pattern Detection Machine learning is used to find patterns within the data. Anomaly and pattern detection are used to determine connections between the information and can determine if the data fits into a pattern or if it is an outlier. This pattern is primarily used to decide which data is similar to other information and which data is different.

Recognition The recognition pattern uses machine learning to specifically identify desired information within unstructured data. Unstructured data is data not easily identifiable such as audio and video. The primary objective of the recognition pattern uses machine learning to identify and understand desired things from unstructured content.

Goal-Driven Systems This uses g machine learning to give people the ability to determine the best solution to a problem An example of a goal-driven system is in a business that needs to find the optimal way to achieve a goal. Using this pattern will allow the business to have the best solutions to possible problems.

Types of Artificial Intelligence

1. Assisted AI: a means of automating simple processes and tasks by harnessing the combined power of Big Data, cloud, and data science to aid in decision-making. Assisted AI is to support humans in making decisions. Assisted AI has the benefit of being used to complete basic tasks, thus freeing up the user to perform more complex tasks.

2. Augmented AI: allows organizations and people to do things they could not otherwise do by supporting human decisions, not by simulating independent intelligence. Augmented AI is more advanced than Assisted AI because Augmented AI can make some decisions on its own but is not completely independent of the user. Overall, Augmented AI suggests new solutions rather than simply identifying patterns and applying predetermined solutions.

3. Autonomous AI: The most advanced form of AI is Autonomous AI, "in which processes are automated to generate the intelligence that allows machines, bots and systems to act on their own, independent of human intervention". Autonomous AI is the most sophisticated type and has the capability to operate without any user interference. Autonomous AI is able to "adapt to their environments and perform tasks that would have been previously unsafe or impossible for a human to do (e.g., the use of drones to perform inventory inspections autonomously of assets in remote locations auditors do not have access to)."

7.2 Data Analytics (Descriptive, Diagnostic, Predictive, and Prescriptive)

Predictive analytics is a subset of data analytics. Predictive analytics can be viewed as helping the accountant or auditor in understanding the future and provides foresight by identifying patterns in historical data. One of the most common applications of predictive analytics in the field of accounting is the computation of a credit score to indicate the likelihood of timely future credit payments. This predictive analytics tool can be used to predict an accounts receivable balance at a certain date and to estimate a collection period for each customer.

7.3   Machine Learning Application

Machine learning is a subset of artificial intelligence that automates analytical model building. Machine learning uses these models to perform data analysis in order to understand patterns and make predictions. The machines are programmed to use an iterative approach to learn from the analyzed data, making the learning automated and continuous; as the machine is exposed to increasing amounts of data, robust patterns are recognized, and the feedback is used to alter actions.

Machine learning is a key subset of artificial intelligence (AI), which originated with the idea that machines could be taught to learn in ways similar to how humans learn. Common examples of machine learning can be found in e-mail spam filters and credit monitoring software, as well as the news feed and targeted advertising functions of technology companies such as Facebook and Google.

In machine learning applications, the expectation is that the algorithm will learn from the data provided, in a manner that is similar to how a human being learns from data. A classic application of machine learning tools is pattern recognition.

Facial recognition machine learning software has been developed such that a machine learning algorithm can look at pictures of men and women and be able to identify those features that are male driven from those that are female driven (Alles et al., 2002).

The machines are programmed to use an iterative approach to learn from the analyzed data, making the learning an automated and continuous process.

The goal of machine learning is to write an algorithm that can be trained using test data to look for specific patterns. For example, if a machine learning algorithm that could look at pictures of animals and identify those that contain cats is desired, one starts by identifying the general characteristics of a cat (four legged, furry animal with a tail) and providing it with a sample set of pictures of animals. Initially, the algorithm would be able to identify animals that do not contain the common characteristics such as snakes (no fur, no legs), birds (no fur, not four legged), and fish (no fur, no legs). But it would need to learn that there are other characteristics (distinctive sounds, claws, body shape) to differentiate it from other four-legged furry animals with tails (Verma & Mani, 2015).

Machine learning and traditional statistical analysis are similar in many aspects. The statistical analysis is based on probability theory and probability distributions, while the machine learning is designed to find the optimal combination of mathematical equations that best predicts an outcome. Thus, machine learning is well suited for a broad range of problems that involve classification, linear regression, and cluster analysis.

In 2011, Johan Perols from the University of San Diego compared six of the most popular statistical and machine learning models being used to detect fraud and determined an overlap in six of the 42 predictors that were consistently chosen by the programs. These predictors included auditor turnover, total discretionary accruals, unexpected jumps in employee productivity, and others that some auditors may have not noticed.

Even if they did notice these irregularities in certain factors within the firm, being able to put these facts into a machine learning software and compare the firm's behavior to other companies helps determine the possible presence of fraud faster and more efficiently than if the auditor were to cross-reference all of these facts themselves.

The predictive reliability of machine learning applications is dependent on the quality of the historical data that has been fed to the machine. New and unforeseen events may create invalid results if they are left unidentified or inappropriately weighted. As a result, human biases can influence the use of machine learning.

Such biases can affect which data sets are chosen for training the AI application, the methods chosen for the process, and the interpretation of the output. Finally, although machine learning technology has great potential, its models are still currently limited by many factors, including data storage and retrieval, processing power, algorithmic modeling assumptions, and human errors and judgment.

Jon Raphael, chief innovation officer at Deloitte, expects machine learning to significantly change the way audits are performed, as it enables auditors to largely "avoid the tradeoff between speed and quality."

Rather than relying primarily on representative sampling techniques, machine learning algorithms can provide firms with opportunities to review an entire population for anomalies. When audit teams can work on the entire data population, they can perform their tests in a more directed and intentional manner. In addition, machine learning algorithms can "learn" from auditors' conclusions on specific items and apply the same logic to other items with similar characteristics.

One of the most tedious and time-consuming parts of an audit is the time it takes to review and extract key terms from contracts. With artificial intelligence, this process has become automated, with systems being taught how to review the same documents and then identify and extract key terms. To solve this problem and speed up the document review process, Deloitte US developed an automated document review platform that natural language processing (NLP) to read electronic documents and machine learning to identify relevant information and flag key terms within the documents.

Machine learning technology for auditing is a very promising area (Dickey, 2019). Several of the Big 4 audit firms have machine learning systems under development, and smaller audit firms are beginning to benefit from improving viability of this technology. It is expected that auditing standards will adapt to take into account the use of machine learning in the audit process. Regulators and standard setters will also need to consider how they can incorporate the impact of this technology in their regulatory and decision-making process. Likewise, educational programs will continue to evolve in this new paradigm. We foresee that more accounting programs with data analytics and machine learning specializations will become the norm rather than the exception.

It is possible that many routine accounting processes will be handled by machine learning algorithms or robotics automation processing (RPA) tools in the near future. For example, it is possible that machine learning algorithms can receive an invoice, match it to a purchase order, determine the expense account to charge and the amount to be paid, and place it in a pool of payments for a human employee to review the documents and release them for payment to the respective vendors.

In the same way, in auditing a client, a well-designed machine learning algorithm could make it easier to detect potential fraudulent transactions in a company's financial statements by training the machine learning algorithm to successfully identify transactions that have characteristics associated with fraudulent activities from bona fide transactions. The evolution of machine learning is thus expected to have a dramatic impact on business, and it is expected that the accounting profession will need to adapt so as to better understand how to utilize such technologies in modifying their ways of working when auditing financial statements of their audit clients.

One example is Deloitte's use of Argus, a machine learning tool that "learns" from every human interaction and leverages advanced machine learning techniques and natural language processing to automatically identify and extract key accounting information from any type of electronic documents such as leases, derivatives contracts, and sales contracts. Argus is programmed with algorithms that allow it to identify key contract terms, as well as trends and outliers. It is highly possible for a well-designed machine to not just read a lease contract, identify key terms, determine whether it is a capital or operating lease, but also to interpret nonstandard leases with significant judgments (e.g., those with unusual asset retirement obligations). This would allow auditors to review and assess larger samples—even up to 100% of the documents, spend more time on judgemental areas and provide greater insights to audit clients, thus improving both the speed and quality of the audit process.

Another example of machine learning technology currently used by PricewaterhouseCoopers is Halo. Halo analyzes journal entries and can identify potentially problematic areas, such as entries with keywords of a questionable nature, entries from unauthorized sources, or an unusually high number of journal entry postings just under authorized limits. Similar to Argus, Halo allows auditors to test 100% of the journal entries and focusing only on the outliers with the highest risk, both the speed and quality of the testing procedures are significantly improved.

In March 2016, Deloitte announced a partnership with Kira Systems to help "free workers from the tedium of reviewing contracts and other documents" (Kira Systems). Kira's advances in machine learning allowed Deloitte professionals to use the technology to simplify complex documents, allowing for quicker analysis.

While originally designed for contracts, Deloitte auditors now use Kira to find "foregone revenues or reduce third party cost and risk" (Kira Systems), and Kira recently released additional platforms for Deloitte's tax and advisory practices.

Expected Innovations in Machine Learning

In May 2018, PricewaterhouseCoopers announced a joint venture with eBravia, a contract analytics software company, to develop machine learning algorithms for contract analysis ("PwC Announces Legal AI partnership with eBrevia for Doc Review," Artificial Lawyer, 2018, https://bit.ly/2APAKZr). Those algorithms could be used to review documents related to lease accounting and revenue recognition standards as well as other business activities, such as mergers and acquisitions, financings, and divestitures. Deloitte has advised retailers on how to enhance customer experience by using machine learning to target products and services based on past buying patterns.

While the major public accounting firms may have the financial resources to invest in machine learning, small public accounting firms have the agility to use pre-built machine learning algorithms to develop expertise through implementations at a smaller scale.

Many routine accounting processes will be handled by machine learning algorithms in the near future. Accounting processes such as expense reports, accounts payable, and risk assessment may be easily automated using machine learning. The jobs requiring the processing of documents have already started disappearing with the advent of document scanners, optical character recognition, and software to match source documents. As an example, machine learning algorithms can receive an invoice, match it to a purchase order, determine the expense account to charge, and place it in a pool of payments to release; a human worker can review the documents and release them for payment. While accounting jobs in businesses will change in the near future, the question of how the public accounting profession will evolve remains (Younis, 2020).

Given that companies will deploy machine learning in their operations to improve accuracy and reduce costs, the advisory services of public accounting firms could dramatically change. It is estimated that 80% of the time spent in advisories services is processing information about a company's operations. Much of this information processing could be handled by machine learning algorithms, meaning that most of the time billed to clients would focus on valued added services that analyze the information produced by machine learning.

The impact of machine learning will most likely be less pervasive in tax preparation services, due to the need for specialized advice and technical research in the context of complex corporate and individual planning issues.

Auditing is an area that will significantly change in the future. Many have predicted that the automation of analyzing a company's financial statements and source documents will result in smaller audit staffs. Auditing standards, however, require an auditor to understand the systems and processes related to the preparation of the financial statements—meaning that the technical expertise required of auditors to understand the machine learning algorithms used in a company's financial systems will be very different from what it is today. Auditors will need to understand the technologies involved and their interaction with internal controls to avoid material misstatements. Potential fraud in a company's financial statements could become easier to identify by using a machine learning algorithm to identify trans-actions that have characteristics associated with fraudulent activities.

Global companies face large and increasingly complicated tax compliance requirements; allocating revenue and expenses to various taxing jurisdictions requires significant data processing and analysis. Machine learning can help tax professionals keep up with relevant tax law changes. Creating algorithms to extract relevant planning information from vast amounts of data is ideal for machine learning. It is hard to accomplish effective tax planning without the relevant and important facts; machine learning can make the fact gathering and analysis function much more efficient and effective. In addition, taxing authorities are exploring the use of machine learning to increase transparency and audit efficiency. The IRS has already begun to develop machine learning algorithms to identify patterns that are associated with tax evasion and fraud. Former IRS agent Michael Sullivan indicated that the public should be aware that the IRS has begun using a new audit method, the 'Machine Learning Tax Audit'. As tax laws continue to grow more complex and the IRS's processes for identifying a taxpayer for an audit become more sophisticated, machine learning may allow tax accountants to better predict deductions that will be disputed by the IRS and identify the regulations that allow for those deductions.

7.4   Data Visualization

Data Visualization is the process of selection, transformation, and presentation of various forms of data in a visual form that helps facilitate exploration and understanding.

The main goal of data visualization is to help users gain better insights, draw better conclusions and eventually generate hypotheses.

This is achieved by integrating the user's perceptual abilities into the data analysis process, and applying their flexibility, creativity, and general knowledge to the large data sets available in today's systems.

In general, there are five main stages to data visualization:

- The collection and storage of data.
- The preprocessing of data.
- The hardware used for display.
- The algorithms used to visualize the data.
- The human perceptual and cognitive system (the process of thinking).

In general, there are two categories of data visualization, each serving different purposes: explanation and exploration. Explanatory data visualization is appropriate when we know what the data is and has to say.

In the accounting and auditing literature, prior research has examined the impor-tance of presentation format and its linkages to decision-making performance. They put focus on the comparison of different visual techniques and their impact on decision-making. Furthermore, the growing number of studies examining presentation format provide an indication of its importance in decision-making.

In contrast, visual data exploration is appropriate when little is known about the data and the exploration goals are vague. Translating large data sets into a visual medium can help in identifying interesting trends and/or outliers. Exploratory data visualization facilitates the user exploring the data, helping them unearth their own insights. Depending on the user's context, it is a discovery process that could or could not potentially lead to the finding of many different insights. Ultimately though, it could help users obtain interesting information, and build hypothesis from large amount of data.

Users who utilize exploratory data visualization generally do not know what the data will show, and would usually analyze and look at the data from a couple of different angles, searching for relationships, connections, and insights that might be concealed in the data. In contrast, users who use explanatory data visualization can typically be called presenters as they are already experts in their own data. They have already explored and analyzed the data and highlighted the data points that support the core ideas they want to communicate.

Specifically, explanatory data visualization is part of a presentation phase, where we want to convey certain information in a visual form. With Big Data, companies need better ways, to not only explore data, but to synthesize meaning from it. Producing visuals that provide explanation and understanding can have significant effects in guiding users toward a conclusion, persuading them to take different actions, or inviting them to ask entirely new questions. Nevertheless, creating such visuals requires preplanning, setting clear objectives, and obtaining the right visual elements.

Data visualization tools are becoming increasingly popular because of the way these tools help users obtain better insights, draw conclusions, and handle large datasets. For example, auditors have begun to use visualizations as a tool to look at multiple accounts over multiple years to detect misstatements.

If an auditor is attempting to examine a company's accounts payable (AP) balances over the last 10 years compared to the industry average, a data visualization tool like PowerBI or Tableau can quickly produce a graph that compares two measures against one dimension. The measures are the quantitative data, which are the company's AP balances versus the industry averages. The dimension is a qualitative categorical variable. The difference between data visualization tools from a simple Excel graph is that this information ("sheet") can be easily formatted and combined with other important information ("other sheets") to create a dashboard where numerous sheets are compiled to provide an overall view that shows the auditor a cohesive audit examination of misstatement risk or anomalies in the company's AP balances. As real-time data is streamed to update the dashboard, auditors could also examine the most current transactions that affect AP balances; thus, enabling the auditor to perform continuous audits. With the real-time quality dashboard that provides real-time alerts, it enables collaboration among the audit team on a real-time continuous basis coupled with real-time supervisory review. Analytical procedures and tests of transactions can be done more continually, and the auditor can investigate unusual fluctuations more promptly. The continuous review can also help to even out the workload of the audit team as the audit team members are kept abreast of the client's business environment and financial performance throughout the financial year.

7.5   Data Mining

Data Mining is a technique which has advanced classification and prediction capabilities and can contribute to fraud detection.

Sun et al. noted that BDA uses data mining to uncover knowledge from a data warehouse or a big dataset to support decision-making creating predictive models to forecast future opportunities and threads, and analyzing and optimizing business processes.

Hence, the Big Data Analytics offer the capability of capturing "sequential causational and correlational processes" on a real-time basis and may change the financial accounting dramatically and reporting that is legacy based, relied on structured data and successive layers of summary and aggregation and reports on a periodic basis.

8   How to Use ADAs in Risk Assessment Procedures?

One common preliminary analytical procedure that can be enhanced through ADAs is performing a year-over-year general ledger fluctuation analysis.

In Step 1 of the ADA process, auditors plan to evaluate all financial statement line items for risk, with the general purpose of assessing the risk of material misstatement for the audited entity's year-end balances. Auditors perform the analysis using year-over-year comparisons in Tableau. Using professional judgment, auditors decide that all accounts that have changed by more than $three million are deemed "notable" and merit additional investigation.

Step 2 of the process is to access and prepare the data. In this example, this process is relatively straightforward, because the data used in this ADA is the trial balance provided to the auditor by management, as well as audited trial balances from the past four years. Because these trial balances exist in standard spreadsheet format, loading them into Tableau™ is a straightforward process.

Step 3 considers the relevance and reliability of the data. The trial balances are clearly relevant as they relate directly to the test of year-over-year differences in account balances. Further, the historical audited data is reliable based on prior year audit results. Because the entity's control environment was effective during interim testing, and because this ADA is a preliminary analytical procedure and not a substantive test, no further tests were performed on the trial balance's data reliability other than agreeing on the beginning balance to the prior year's audited trial balance.

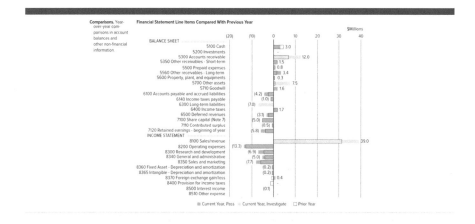

Fig. 1 Audit Data Analytics (ADAs)

Step 4 is to perform the ADA, shown in Fig. 1. Dark blue bars represent year-over-year fluctuations that are below the $three million threshold, while light blue bars are those that exceed the threshold. Accounts Receivable, Other Assets, Long Term Liabilities, and Revenue all changed by amounts exceeding the threshold. Based on previous inquiry, the significant change in the year-end Accounts Receivable (AR) balance was most unexpected. Therefore, in response to this initial ADA, auditors follow up with an additional ADA focusing on AR. Because of the audited entity's multinational customer base, the auditors use a trend analysis to examine AR balance, by currency, over the past 5 years.

9 How to Use Audit Data Analytics (ADA) in Substantive Analytical Procedures?

Audit data analytics used as a substantive analytical procedure often use more data, or different data (for example, more disaggregated), with different techniques than traditional substantive analytical procedures. However, when auditors use Audit Data Analytics (ADAs) as substantive analytical procedures, they must be careful to follow the explicit audit guidance related to substantive analytical procedures including careful documentation of how the auditor developed an expectation, defined a tolerable difference, performed the analytic, and investigated differences.

10 How to Use ADAs in Tests of Detail?

To this point, ADAs have been presented primarily as more advanced analytical procedures and visualizations. However, ADAs can also be used as a test of detail because of their ability to process and analyze large amounts of data. For repetitive processes and calculations, ADAs can be used to test entire populations that have previously been tested using samples. For example, instead of testing a sample of contracts for revenue purposes, software can be used to have the computer "read" *all* contracts and search for predetermined key phrases that indicate side agreements or unusual terms that indicate a heightened risk of material misstatement.

References

Alles, M. G., Kogan, A., & Vasarhelyi, M. A. (2002). Feasibility and economics of continuous assurance. *Auditing: A Journal of Practice & Theory, 21*(1), 125–138.

Appelbaum, D., Kogan, A., & Vasarhelyi, M. (2015). Moving towards continuous audit and big data with audit analytics: Implications for research and practice. *Rutgers Accounting Research Center (RARC) and Continuous Auditing & Reporting Laboratory (CAR-Lab).*

Balios, D., Kotsilaras, P., Eriotis, N., & Vasiliou, D. (2020). Big data, data analytics and external auditing. *Journal of Modern Accounting and Auditing, 16*, 5.

Dickey, G. (2019). Machine learning in Auditing current and future applications. *CPA Journal, 2019*(June), 1–10.

Jesus, M. (2018). *How machine learning is disrupting the accounting industry.* BigML Blog. https://bit.ly/37hIvU5

Rajvanshi, A. (2016). *Predictive Analytics/machine learning for fraud detection.* Business Intelligence Engineer at Amazon.

Verma, R., & Mani, S. R. (2015). Using analytics for insurance fraud detection. *FINsights, Infosys, 10*, 1–23.

Younis, N. M. M. (2020). The impact of big data analytics on improving financial reporting quality. *International Journal of Economics, Business and Accounting Research (IJEBAR) Peer Reviewed – International Journal, 4*(3) E-ISSN: 2614-1280 P-ISSN 2622-4771.

Digitalization in Passengers' Air Travel

Jiezhuoma La ⓘ, Iryna Heiets, and Cees Bil

1 Introduction

Digitalization is a certain trend in the aviation industry. As Abeyratne stated digitalization and digital technology are changing the aviation industry's direction (Abeyratne, 2020). With high efficiency, digital processes and automatic measures are gradually replacing the manual process in aviation and relevant industries (Bushnell, 2020). Indeed, the aviation industry developed rapidly before 2019. Due to economic development, population increase, and technology development, the aviation industry got a good performance in the last decades. However, this good performance was rapidly changed by the flight ban and travel restrictions caused by the COVID-19 pandemic. The aviation industry is suffering a huge loss, and it needs a quick recovery at the post-pandemic stage. Although the global aviation market has already started to take some actions like launching digital travel passes (IATA, 2020), these actions are not enough for the industry to bounce back as before. Thus, it is urgent to study how to use digital technology to improve the aviation industry's performance to a greater extent.

As NASA proposed in 2020, the door-to-door (D2D) model could be an innovative transport method which brought new opportunities to the aviation industry (McGrath, 2002). In recent years, more and more organizations and institutions realize the importance of this model. However, the research on the door-to-door model is just at the beginning stage. Although some airlines and airports start to collaborate with selected land transportation mode, these collaborations still cannot cover the entire travel stage and the efficiency of the process also fails to get a high improvement. Thus, digitalization can be a new measure to strengthen the

J. La (✉) · I. Heiets · C. Bil
RMIT University, Melbourne, VIC, Australia
e-mail: s3633823@student.rmit.edu.au; iryna.heiets@rmit.edu.au; cees.bil@rmit.edu.au

connection between the different transportation modes in the D2D and improve the aviation industry's performance.

This study aims to investigate how digital technology impacts passengers' D2D air travel from a passenger perspective. Survey research was conducted to collect data and analyze how passengers evaluated the digital technologies which have been applied in their entire travel process. Then, both the academics and the industry can get a better understanding of how to use digital technology to improve the D2D model and to satisfy their customers.

There are four parts in this study. Firstly, the previous studies about digitalization and aviation will be reviewed. Then, the research method and relevant data analysis method will be introduced. Besides, the main part of this study is the results and analyses part, which will be provided with a detailed discussion. At last, some recommendations and a conclusion will be provided to further the results, and the research plan for the next step will also be briefly introduced.

2   Literature Review

2.1   Digitalization and Digital Technology

Digitalization is a new concept which changes the development trend of both society and industries (Gray & Rumpe, 2015). Digital technology and digitalization can provide industries and businesses with an opportunity to generate more revenue. Besides, it also can help the markets get the right balance between supply and demand (Gray & Rumpe, 2015). Actually, digitalization and digital technology can not only be applied to improve the production process but also can clear the relationship between businesses and customers (Gray & Rumpe, 2015). In the past, there was a blurry line between the responsibility of industry and the responsibility of customers. However, digital technology helps the stakeholders to clarify their roles and responsibility and brings a win-win to both the businesses and customers. As Parida asserted in 2018, digitalization and digital technology can rebuild business models and improve the industries' strategic ability (Parida, 2018). Digitalization is a force which has been created by Industry 4.0 and the Internet, and it changes the way people think. It has been confirmed that digitalization leads to a huge transformation in society and the industry (Parida, 2018).

As a cutting-edge technology, digital technology starts to be accepted by the industry and market gradually. More and more business leaders take the value of digitalization and try to enter the digital market earlier to win the monopolistic place (Parida, 2018). Actually, higher educational institutions, technology companies, and professional associations are all the relevant stakeholders that need to involve digitalization in their businesses and develop their own digital strategies (Crittenden et al., 2019). However, it is not so easy to achieve the digital transformation successfully without any professional guidance. Parviainen et al. proposed a new model to help enterprises develop their digital strategy and achieve the digital transformation easier (Parviainen et al., 2017). The core of this model is to identify the current digital state of the business. Then, the most suitable digital strategy could be developed for the organizations to get the improvement. In this situation, in order to help the industry to improve their performance rapidly, the changes that digitalization and digital technology have brought to the industries, especially the aviation industry, should be investigated.

2.2 Current Aviation Industry

The outbreak of COVID-19 has a great impact on the global aviation industry. Air travel is one of the important reasons for the spread of Coronavirus in the global area (Daon et al., 2020). Since the outbreak of the epidemic at the end of 2019, various countries have issued corresponding policies to restrict air travel. These restrictions have caused a sharp drop in flight volume, which is the contrast with the previous forecasts for the aviation industry (Daon et al., 2020). The long-term flight ban has affected multiple sectors of the global economic system. Besides, the huge reductions in ticketing and passenger numbers have caused a sharp decline in aviation revenue on a global scale. In this situation, numbers of countries start paying attention to launch relevant policies and using different technologies to mitigate the impact of epidemic (Iacus et al., 2020).

The impact of COVID-19 on different types of airlines is also different. Suau-Sanchez et al. have conducted an interview with some representatives from the aviation industry in 2020 (Suau-Sanchez et al., 2020). When talking about full-service airlines, most interviewees believed that the COVID-19 pandemic had a greater impact on this kind of airline companies. The reduced number of passengers and increased competition in the airline industry have greatly weakened the profits of full-service airlines. It also increased the difficulties in the full-service airlines' operation. For low-cost airlines, the interviewees agreed that most low-cost airlines would transfer their businesses from small markets to some large markets. Reducing the flight frequencies would also be a possible measure for low-cost airlines to maintain operation. All in all, the reduction of demand in the aviation market has a direct impact on the business and hierarchy of the aviation industry (Suau-Sanchez et al., 2020). Thus, more digital applications need to be applied to help the industry overcome the negative impact.

2.3 Digital and Aviation Industry

In the current modern era, both the civil aviation market and the military aviation market have applied various digital technologies in their operation and management. Artificial intelligent technology, biometric technology, big data, cloud, and many other digital technologies have been gradually used in the aviation industry (Chang et al.,

2019). These digital tools are effective and useful, enabling the aviation industry to become a major player in the global travel market. In addition, digitalization also provides the aviation industry with a good opportunity to maintain an outstanding position in the competitive market. With so many benefits, the application of digital technology should be promoted in the industry and the gap between academia and industry should be resolved (Chang et al., 2019). Although the application of digital technology in the aviation industry is still at the beginning stage, the digital competition in the aviation industry is already intensive (Çakır & Ulukan, 2022). Airlines and airports need more advanced systems to improve their working efficiency and to achieve digitalization easier. Besides, since the aviation industry needs passengers' comments to develop a practical strategy, these advanced technologies also help the aviation industry get a chance to directly contact their customers to get useful feedback (Çakır & Ulukan, 2022).

Digitalization features a constantly developing concept and it is also recognized as one of the most innovative methods to help the aviation industry develop at the current process (Kuisma, 2018). Besides, the stakeholders in the aviation industry are highly recommended to develop a solid digital strategy, so that they can fit into the competitive and dynamic market faster and easier (Kuisma, 2018). As the aviation industry must compete with both their competitors and customer's satisfaction, it should be the first priority to develop some suitable digital tools to serve the industry (Çakır & Ulukan, 2022). The aviation industry keeps identifying the extent of passengers' satisfaction with current digital applications in their trip because they believe they can clarify the future development trend based on the passengers' demands. Thus, it is worth investigating how passengers evaluate the current digital tools applied in their whole trip.

2.4   Digital and Passenger

Passenger is one of the main elements in the aviation industry. Thus, it is worth identifying how passenger satisfaction could be improved (Nurhadi et al., 2019). Currently, air travel passengers can experience self-check-in, e-passport, online ticketing, and many other digital tools in their trip. Besides, e-commerce and inflight entertainment system are the digital products that passengers get in contact with more frequently (Hanke, 2016). Since these two digital technologies have been applied in the market earlier, passengers are more familiar with them. Thus, compared with self-check-in and digital boarding pass, these two digital tools are ranked at higher positions than passengers satisfy with (Nurhadi et al., 2019). In this situation, there is a question about how to examine the relationship between the different digital tools and passenger satisfaction. A customer survey will be a good answer. Since surveys can directly get feedbacks from real passengers, it can be an effective method to help the aviation industry to make the digital plan via using the collected information (Nurhadi et al., 2019). According to the survey conducted by Nurhadi in 2019, passengers were satisfied with the inflight digital tools rather than the preflight digital tools. This survey could provide some useful information about the extent of passengers' satisfaction with current digital tools. However, it only covers the pre- and in-flight stages, which is inadequate to identify passengers' real needs and wants in their whole trip. Thus, relevant research about how passengers evaluate digital technologies in their D2D air travel process should be taken.

2.5 Passenger and Door-to-Door (D2D) Model

Travel time, budget, experience, and value are the main elements that passengers focus on. With the increasing customer demands and the COVID-19 impacts, the aviation industry starts to pay more attention to passengers' entire travel process rather than the segmented parts (Kluge et al., 2020). Thus, a relevant D2D model has been proposed based on passengers' needs. D2D air travel model is a holistic model that covers all passengers' travel stages, and it gives passengers an opportunity to manage their whole trip in one model (Kluge et al., 2020). Besides, the D2D model is an innovative and unique model that can enhance the aviation industry's competitive capability (Kluge et al., 2019). By integrating the different travel stages, D2D helps passengers enjoy a seamless and smooth air travel process. Besides, digitalization will be the main driving force for the development of door-to-door air travel (Kluge et al., 2020). Thus, there is a need to analyze how digitalization and digital technology impact passengers' D2D process.

3 Research Method

3.1 Survey Research

The main research method in this study is the survey research. As introduced in the Sect. 2.4, survey research is an effective method that is used to collect passengers' real opinions about the current digital technologies in the aviation industry. Survey research is a systemic method that can use for data collection from a specified group of population (Schwarz et al., 1999). This research method can help the researchers collect the responses directly from their target populations (Fowler Jr, 2013). The main parts of survey research contain objective identification, question design, approach design, population selection, sample design, data collection, and adjustment and data analyses. All these processes should be designed scientifically and reasonably. In this study, survey research was conducted to collect passengers' ideas about how they evaluated the different digital tools in their D2D journey and how they evaluated the sustainability of digitalization and digital technology in D2D air travel process.

The survey was conducted via online platforms due to the current COVID-19 pandemic situation. Both quantitative questions and qualitative questions were designed in the survey. Besides, in order to simplify the process and ease the data analysis, all questions were designed based on the different travel stages in passengers' D2D trip. A total of 220 persons aged 20 and older were surveyed in this study.

3.2 Data Analyses

All data used and analyzed in this study were collected from the survey research. Cross-table analysis and frequency analysis were the main methods used to calculate and analyze the collected data. In this research, these two methods were used to analyze how different types of passengers evaluate digital technologies in their entire travel process. After this activity, both academia and the industry can get a better understanding of how to use digital technology to improve passenger satisfaction. Besides, the results could also help the relevant stakeholders to establish a more practical and effective digital strategy.

4 Results and Discussion

This section provides the data analysis results and relevant discussions. All participants involved in this survey were voluntary and confidential. Before starting the survey, all participants could go through the questions and decided if they were willing to take part in this activity. Besides, some participants might fail to answer some questions, and this kind of response was marked as "missing value." There are totally 220 responses to this survey.

4.1 General Journey

This section contains information about how passengers evaluated digital tools in their general journey. Table 1 gives information about how often that different travel purposes passengers used digital tools in their air travel. This question was about passengers' general travel process, and all currently available digital technologies were included in it. After analysis, decision makers can get a primary understanding of how to improve the travel experience for different types of passengers.

Table 1 shows the analysis results about the use frequency of digital tools in passengers' general air travel process.

Table 1 shows the percentages of different passengers using digital tools during their air travel. In this question, passengers' travel purposes can be classified as leisure, business, visiting friends and family, and others. It can be found that more than half of the passengers used digital tools every time when they were traveling. In this group, the number of passengers who traveled to visit their family and friends

Table 1 Cross-table: Frequency of using digital tools in general air travel process

Frequency	Travel purpose				
	Total	Leisure	Business/Work	Visiting friends and family	Other reasons
Every time	56.4%	59.9%	46.7%	61.5%	25.0%
Almost	19.5%	18.4%	30.0%	15.4%	16.7%
Sometime	14.1%	12.5%	20.0%	11.5%	25.0%
Rarely	5.5%	5.9%	3.3%	7.8%	0.0%
Never	4.5%	3.3%	0.0%	3.8%	33.3%
Total	100.0%				

Fig. 1 The percentage of digital tools that participants most likely to use

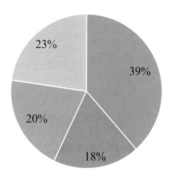

- Airline Apps/ Websites
- Digital Facilities in the airport
- Digital facilities in flight
- Websites or social media of airlines and/or airports

occupied the largest position, which was 61.5% of this kind of passengers. Besides, the using frequency as almost was ranked in second place in Table 1. Different from the "every time" group, the number of business/work purpose passengers occupied the largest place in the "almost" group as 30.0%. Besides, the group of passengers who never used digital tools before in their travel process occupied the smallest part as 4.5% of total passengers. In this group, the number of passengers who traveled for other reasons was largest, which was 33.3% of this kind of passengers. As analyzed above, it can be found that business/work purpose participants took the highest value of digital technology. Moreover, most of the participants have used digital tools in their air travel process before.

Figure 1 gives information about what kinds of digital tools passengers were most likely to use in their general journey. These digital tools were classified into four categories: airline apps, airport facilities, inflight facilities, and social media of airlines/airports. Participants could choose more than one answer to this question.

Figure 1 shows the percentages of digital tools which passengers used most commonly in their general trips. It can be seen from the figure that airline apps and websites occupied the largest place, which was 39.0%. Besides, airlines/airports' social media was the second most commonly used digital tool in passengers' general trips. Thus, it can be concluded that more than half of the participants liked to use online platforms to get the service they needed during their air travel. Moreover, the inflight digital tool was ranked as the third one and the airport digital facility was ranked as the last one in this question, which were 20.0% and 18.0%, respectively. This result also supports the analyses in Sect. 2.4. Thus, it can be concluded that the maturity and early start of digital tools will be the main factors that affect passengers' using experience.

As analyzed above, more than half of the passengers were willing to use or already used digital tools in their general trip, and the using frequency of these digital technologies was also high. Among these passengers, a large part of participants attached more importance to online ticketing or social media since these digital technologies were more mature and started earlier. Thus, it is worth investigating the impacts of the digital tools which have been applied in the different travel stages. Based on this, some relevant guidelines can be developed to improve passengers' satisfaction.

4.2 Stage 1—From Origin to Airport and Stage 5—From Airport to Destination

This section introduces the relationship between digital technology with passengers at the stage 1, which is the stage of passengers departing from their origins to the airports, and the stage 5, which is the stage of passengers departing from the airports to their destinations. These two stages are the first stage and the last stage in the D2D model. Questions in these stages mainly focused on the online ticketing and online land transport service. The repossess to these questions were provided as "agree or disagree."

Table 2 shows the information on how passengers evaluated the current online ticketing platforms. These online platforms included both airlines' apps/ websites and travel agencies' apps/websites. Besides, passengers were classified by their travel frequency in this table.

Table 2 Cross-table: How passengers evaluate online ticketing platforms

Willing to purchase flight tickets via online platforms	Travel frequency					
	Total	None	1–2 times	3–5 times	6–10 times	More than 10 times
Strongly agree	72.2%	61.0%	71.1%	76.5%	83.3%	78.3%
Agree	23.0%	27.8%	25.8%	17.6%	16.7%	13.0%
Neutral	2.4%	5.6%	0.8%	3.0%	0.0%	8.7%
Disagree	2.4%	5.6%	2.3%	2.9%	0.0%	0.0%
Strongly disagree	0.0%	0.0%	0.0%	0.0%	0.0%	0.0%
Total	100.0%					

Table 3 Cross-table: Passengers' concerns of the security issues with online ticketing

If passengers concerned about security issues while purchasing ticket online	Online ticketing would be cheaper than purchasing tickets over the counter.					
	Total	Strongly agree	Agree	Neutral	Disagree	Strongly disagree
Strongly agree	25.4%	34.4%	16.1%	22.6%	0.0%	0.0%
Agree	32.1%	16.7%	50.0%	34.0%	75.0%	0.0%
Neutral	25.3%	28.9%	21.0%	26.4%	0.0%	0.0%
Disagree	13.9%	14.4%	11.3%	15.1%	25.0%	0.0%
Strongly disagree	3.3%	5.6%	1.6%	1.9%	0.0%	0.0%
Total	100.0%					0.0%

According to Table 2, almost all of the participants thought they were able to purchase their flight tickets via online platforms. Besides, the number of passengers who were willing to purchase online occupied the largest position, which was 72.2%. When looking at passengers' frequency, participants who traveled 6–10 time a year took the highest value of online ticketing, and passengers who traveled more than 10 times a year was ranked as the second. In contrast, except for the participants who never traveled by air before, passengers who travel 1–2 times a year took the lowest value of online ticketing. Thus, it can be found that passengers' attitude toward online ticketing is positively related to passengers' air travel frequency.

Table 3 shows the information on how different cost-sensitive participants evaluated the security issues with the current online ticketing tools.

It can be found in Table 3, for passengers who did not think online ticketing was cheaper than other ticketing channels, they also did not worry about the security issues within the online purchasing process. Besides, passengers who thought online ticketing was cheaper contributed to the main part of passengers who strongly worried about the security issues. Moreover, for participants who asserted a natural attitude toward the online ticketing cost, they also held the same attitude toward the security issues. Thus, this result illustrated that passengers would show more concern to the potential security issues in the online ticketing process if the online platforms provided cheaper tickets than other channels.

Currently, there are some online services provided by airlines and airports to help passengers book land transportation and connect the different transport modes easily, e.g., land transportation from passengers' origins to departure airports and land transportation from the arrival airport to passengers' destinations. Table 4 shows the counts and percentages about if passengers thought they could save more time by using online airlines' chauffeur service.

According to Table 4, more than 52.8% of participants believed that the online airlines' chauffeur service could help them to save their land transport time by simplifying the booking and ticket purchasing processes. Besides, there was also a large part of passengers who asserted a natural attitude toward this online service since they thought the online airlines' chauffeur service performed not so well as it

Table 4 How passengers evaluated the online airlines' chauffeur service

Category	Count	%
Strongly agree	34	15.5%
Agree	82	37.3%
Neutral	78	35.5%
Disagree	8	3.6%
Strongly disagree	6	2.7%
Missing value	12	5.5%
Total	220	100.0%

Table 5 Cross-table: Passengers' concerns of the security issues with online service

If passengers concerned about security issues while using this service	If used the online airlines' chauffeur services before		
	Total	No	Yes
Strongly agree	15.5%	32.1%	15.8%
Agree	36.7%	25.0%	61.4%
Neutral	29.5%	21.4%	17.5%
Disagree	14.0%	14.3%	5.3%
Strongly disagree	4.3%	7.2%	0.0%
Total	100.0%		

claimed. Moreover, except for the missing value, only 6.3% of participants said this online service was totally unhelpful in helping passengers save their land transport time. Thus, it can be summarized that most of the passengers thought this online service could help passengers to improve their travel efficiency at stages 1 and 5.

Table 5 shows the information about the extent of passengers worried about the security issues in the online airlines' chauffeur service. This table contains responses from both former and non-former users of this service.

Table 5 shows the percentages of passengers who are worried about the security issues with the online airlines' chauffeur service. It can be found that most of the former users showed their concerns to the potential security issues when they were using this online land transport service. Besides, the number of non-former users who worried about the security issues in this service was not as much as the number of former users. However, there was still more than 57.1% of non-former users who held a negative attitude toward this service. Thus, the security issues in using this online service need to be resolved to attract more users.

As analyzed above, in this stage, most of the participants thought digital technology could bring positive impacts on their travel process. However, the potential security issues within these online processes should be addressed effectively to improve passengers' confidence in using these digital tools.

4.3 Stages 2 and 4—At Airports

This section introduces how passengers evaluated the digital technologies and digital tools applied in both the departure airports and the arrival airports. Questions in these two stages mainly focused on the check-in, luggage handling, and customs processes. Besides, the responses to these questions were provided as "yes or no" or "agree or disagree."

Table 6 shows the information about how different travel purposes passengers evaluated the self-check-in kiosks and automatic luggage handling facilities.

As the results are shown in Table 6, most of the participants agreed or strongly agreed that the digital facilities in the airports could help them to save more time in stages 2 and 4. As about 87.4% of participants said they thought these digital tools could improve their travel efficiency, passengers' attitude toward digital technology in airports was positive. Among these participants, the percentages of the business purpose passengers and the leisure purpose passengers occupied the largest places, which were 51.4% and 50.0%, respectively. Besides, there were also another 4.4% of participants who were highly unsatisfied with these digital tools. Although this number was not huge, more actions should be taken to improve passenger satisfaction and maintain passengers' loyalty to the industry.

Some airlines provide service that enables passengers to do the online check-in for their flights and online check-in for the booked hotels at their destinations at the same time. This is also a useful connection between passengers' different travel stages, and it can improve the efficiency of the entire travel process. Figure 2 shows the percentage of participants who thought this service was applicable.

More than 73% of passengers thought the service which connected the flight online check-in process and the hotel online check-in process together was helpful and applicable. These participants believed the service could be a useful measure to improve travel efficiency. However, there were another 26.2% of passengers worried about the reliability of this service. They worried that if delays or changes happened in the journey, this service might bring difficulties to passengers to change their following trips. Thus, more improvements are needed in this service to take disruption and delay situations into mind and improve the process.

Table 6 Cross-table: How passengers evaluated the self-check-in kiosks and automatic luggage handling facilities

If these digital tools can help to save more time	Travel purpose				
	Total	Leisure	Business / Work	Visiting friends and family	Other reasons
Strongly agree	49.8%	51.4%	50.0%	43.5%	40.0%
Agree	37.6%	36.1%	39.3%	43.5%	40.0%
Neutral	8.2%	8.3%	3.6%	13.0%	10.0%
Disagree	2.9%	2.1%	7.1%	0.0%	10.0%
Strongly disagree	1.5%	2.1%	0.0%	0.0%	0.0%
Total	100.0%				

Fig. 2 The percentage of
passengers' attitude

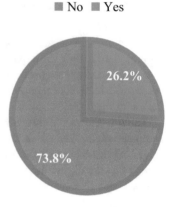

■ No ■ Yes

26.2%

73.8%

Table 7 Cross-table: How passengers evaluated the automatic immigration facilities

Can the automated immigration facilities help passengers save time	If used the automated immigration facilities before		
	Total	Yes	No
Strongly agree	47.5%	57.4%	5.3%
Agree	33.2%	32.3%	36.8%
Neutral	17.8%	8.5%	57.9%
Disagree	1.0%	1.2%	0.0%
Strongly disagree	0.5%	0.6%	0.0%
Total	100.0%		

Currently, e-passport has gradually been accepted by passengers. Besides, the number of countries and airports which have applied this technology is increasing. Since passengers' dissatisfaction always occurs in the customs process, improving passengers' experience in this process can be an effective method to help passengers enjoy a more satisfying journey. Table 7 shows information about how participants evaluated the efficiency of e-passport and relevant technologies.

Table 7 gives information about how former users and non-former users thought about the digital technologies in the current customs process. It can be seen from the table, most of the former users uttered that the automatic immigration facilities could improve the efficiency of the customs process. However, the situation was different for the non-former users. For passengers who did not use these technologies before, they believed the automatic immigration facilities could not help them to improve travel efficiency. This would also be the reason why these participants refused to use these digital tools in their air travel. In this situation, more promotions should be done to introduce these technologies to passengers.

Based on the above analyses, most former users held a positive attitude toward automatic immigration facilities. Thus, in the next step, there is a need to investigate the reliability of these technologies. In Table 8, it shows passengers' concerns about

Table 8 Cross-table: Passengers concerns of the security issues with the automatic immigration facilities

If passengers concerned about security issues while using this service	Air travel frequency					
	Total	None	1–2 times	3–5 times	6–10 times	More than 10 times
Strongly agree	7.4%	18.8%	6.4%	9.0%	16.7%	0.0%
Agree	17.3%	18.8%	20.8%	15.2%	16.7%	0.0%
Neutral	29.8%	25.0%	32.8%	27.3%	33.3%	18.2%
Disagree	26.7%	31.2%	22.4%	33.3%	16.7%	40.9%
Strongly disagree	18.8%	6.2%	17.6%	15.2%	16.6%	40.9%
Total	100.0%					

the security issues with the automatic immigration facilities based on passengers' travel frequency.

According to Table 8, it can be found that the number of participants who thought the automatic immigration facilities were reliable occupied the largest place. These kinds of passengers also did not worry about the potential security issues in this process. It was interesting that compared with passengers who traveled less than 10 times per year, passengers who traveled more than 10 times a year had more confidence in these digital tools. Besides, most of the participants who took flights 6–10 times per year showed natural or negative attitudes toward the digital technologies in the customs process. Thus, although a large part of the participants supported the reliability of current automatic immigration facilities, the opponents' opinions should also be considered to upgrade these technologies.

As analyzed above, although a lot of participants thought the digital technologies applied in stages 2 and 4 were reliable and effective, the number of participants who asserted the opposite idea should not be omitted.

4.4 Stage 3—Inflight

Passengers prefer to have a comfortable inflight experience. With the technology developed, more passengers show their interest to use digital tools to spend their inflight time. This section contains information about how participants evaluated the current digital tools in their inflight stage. Questions in this stage mainly focused on the inflight entertainment systems. Besides, the responses to these questions were provided as "agree or disagree."

Table 9 shows the responses to the question that if the inflight systems could be more helpful for passengers to take activities like ordering meals or doing online shopping. By analyzing this table, the efficiency of current inflight entertainment systems could be obtained.

It can be seen from Table 9, about 80% of participants deemed that inflight entertainment systems could effectively support different inflight activities. Besides,

Table 9 Cross-table: How passengers evaluated the inflight entertainment systems

The inflight entertainment systems are efficient enough	Travel purpose				
	Total	Leisure	Business / Work	Visiting friends and family	Other reasons
Strongly agree	39.5%	41.4%	33.3%	39.2%	30.0%
Agree	40.0%	37.9%	44.5%	47.8%	40.0%
Neutral	13.5%	12.9%	11.1%	13.0%	30.0%
Disagree	6.5%	7.1%	11.1%	0.0%	0.0%
Strongly disagree	0.5%	0.7%	0.0%	0.0%	0.0%
Total	100.0%				

passengers who traveled for leisure and passengers who traveled for visiting friends and family took the highest value of these digital systems. The group of business/ work purpose passengers was ranked at third place, in this group, only 22.2% of passengers thought these digital systems were ineffective. Besides, when looking at passengers who traveled for other reasons, the number of these participants who agreed with these in-flight systems' high efficiency was the smallest. Thus, in order to improve passenger satisfaction with inflight digital tools like inflight entertainment systems, more functions should be developed in these systems to serve passengers with their inflight activities.

4.5 Sustainability

Sustainability is an important element for both the industry and the society. Eco-friendly and sustainable development have been mentioned frequently on academic platforms and social media. Thus, it is important to determine if the digitalization happened in passengers' D2D travel is sustainable and durable for the aviation industry and environment development.

Figure 3 gives the percentages of different participants who agreed or disagreed that digital technologies applied in the D2D could bring benefits to both environment and the aviation industry.

From the bar chart shown above, it can be seen that about 76.4% of participants believed that the current digital tools and applications used in the aviation industry were eco-friendly, meanwhile, the percentage of participants who thought digitalization could bring sustainable development to the aviation industry was also 76.4%. Moreover, the percentages of participants who held a natural attitude toward digitalization's sustainability in the environment and in the aviation industry were 12.3% and 13.6%, respectively. Besides, the percentages of participants who thought digitalization and digital technology could not bring sustainable development to the environment and the aviation industry were both less than 2.0%. Indeed, nearly 80% of the participants uttered that digitalization positively impacted the industry

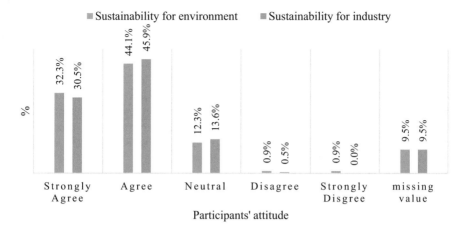

Fig. 3 Participants who thought digital technologies in aviation were sustainable

and environment development. Furthermore, digitalization can lead to a situation where more manual processes will be replaced by artificial intelligence and digital processes. Thus, it will be both a challenge and an opportunity to apply advanced digital technologies to generate more green energies in the industry and the society.

5   Conclusion and Implication

5.1   Conclusion

According to the analyses above, digitalization plays a vital role in passengers' travel process. Passengers were willing to use more useful digital tools in their general trips to manage their travel process. Besides, as Nurhadi et al. stated, online ticketing and online services were two digital tools that passengers attached more importance to in the D2D stage 1 and stage 5 (Nurhadi et al., 2019). Moreover, the number of passengers who used digital facilities in airports was increasing. As mentioned in Sect. 4.3, more than half of the participants deemed that the current digital tools in the aviation industry were useful and effective. However, the number of passengers who were unsatisfied with these digital technologies could not be ignored. One of the reasons that caused the unsatisfaction was that these passengers were unfamiliar with these digital technologies, and they even did not know how to use these high-tech facilities. Thus, more promotions and instructions should be provided to passengers to help them use the digital facilities easier. Furthermore, there are also about 20.0% of passengers who thought the functions of the current inflight entertainment systems were inadequate. Passengers always desire to make good use of their inflight time, however, the functions available in the current inflight digital systems cannot satisfy passengers' needs. Thus, more functions, like real-time flight information and WIFI connection, should be provided in the inflight systems to help passengers experience a more effective travel experience.

Digitalization and digital technology can generate deep impacts on both the aviation industry and the environment. Besides, the digital trend in the global market is recognized as a force that can push the industry to transfer from the conventional operation process to an innovative and high-tech process. However, the pandemic of COVID-19 caused a huge negative impact on the aviation industry's operation. In this situation, it is significant to analyze how the industry can reduce losses via digitalization. One possible suggestion is to increase the implementation of digital tools to help the aviation industry enhance its competitive capability and get a better performance. In addition, some new digital products, such as digital travel passes and inflight service tools, could be developed and applied to attract more target customers and enhance passenger satisfaction.

5.2   Research Limitation and Future Work

The first limitation in this study is time limitation. Due to the limited time, the responses collected in the survey research were only 220. Thus, more research activities are needed in this field to strengthen the result. Besides, due to the COVID-19 pandemic, this research could only be conducted via online platform. Thus, in the next step, a survey about how passengers evaluate digital air travel should be conducted in the airports to receive more reliable responses. Moreover, there is also a scope limitation in this study. In the question design stage, only a part of digital technologies have been involved in the survey. Thus, more up-to-date technologies should be included in future research to identify the extent of passenger satisfaction with these technologies. All in all, due to the limitation of data collection and the special epidemic situation, more samples and resources should be analyzed to strengthen the results of this study in the future.

In the next stage, a new digital door-to-door model, which enables passengers to manage their entire travel process, will be developed to bring a more effective and smooth air travel experience to passengers, and then passengers' loyalty to the aviation industry can also be improved.

Acknowledgments The authors would like to express gratitude to Surein S/O Balachanthar, Pritpal Singh s/o Satwant Singh, Valerie Chua Zi Rui, and Noreffendy Bin Nordzi for their help in survey data collection. This study cannot be completed without their help and support.

References

Abeyratne, R. (2020). *Aviation in the Digital age*. Springer International Publishing.
Bushnell, D. M. (2020). The coming digital reality. *Aerospace Americana, 58*(6), 22–23.

Çakır, E., & Ulukan, Z. (2022). Digitalization on Aviation 4.0: Designing a Scikit-Fuzzy control system for in-flight catering customer satisfaction. In *Intelligent and Fuzzy Techniques in Aviation 4.0* (pp. 123–146). Springer.

Chang, S., Wang, Z., Wang, Y., Tang, J., & Jiang, X. (2019). Enabling technologies and platforms to aid digitalization of commercial aviation support, maintenance and health management. In *2019 International Conference on Quality, Reliability, Risk, Maintenance, and Safety Engineering (QR2MSE)* (pp. 926–932). IEEE.

Crittenden, W. F., Biel, I. K., & Lovely, W. A., III. (2019). Embracing digitalization: Student learning and new technologies. *Journal of Marketing Education, 41*(1), 5–14.

Daon, Y., Thompson, R. N., & Obolski, U. (2020). Estimating COVID-19 outbreak risk through air travel. *Journal of Travel Medicine, 27*(5), taaa093.

Fowler, F. J., Jr. (2013). *Survey research methods.* Sage Publications.

Gray, J., & Rumpe, B. (2015). Models for digitalization. *Software & Systems Modeling, 14*(4), 1319–1320.

Hanke, M. (2016). Airline e-Commerce: Log on. In *Take off* (Vol. 19). Routledge.

Iacus, S. M., Natale, F., Santamaria, C., Spyratos, S., & Vespe, M. (2020). Estimating and projecting air passenger traffic during the COVID-19 coronavirus outbreak and its socio-economic impact. *Safety Science, 129*, 104791.

IATA. (2020). *Economic Performance of the Airline Industry.* IATA.

Kluge, U., Paul, A., Urban, M., & Ureta, H. (2019). Assessment of passenger requirements along the door-to-door travel chain. In *Towards user-centric transport in Europe* (pp. 255–276). Springer.

Kluge, U., Ringbeck, J., & Spinler, S. (2020). Door-to-door travel in 2035–a Delphi study. *Technological Forecasting and Social Change, 157*, 120096.

Kuisma, N.: Digitalization and its impact on commercial aviation (2018).

McGrath, R. N. (2002). A study of NASA's vision for the future of air travel. *Technological Forecasting and Social Change, 2*(69), 173–193.

Nurhadi, M. I., Ratnayake, N., & Fachira, I. (2019). The practice of digitalization in improving customer experience of Indonesian commercial aviation industry. In *The 4th ICMEM 2019 and The 11th IICIES 2019, August, 7–9.*

Parida, V.. (2018). Digitalization. pp. 23–38.

Parviainen, P., Tihinen, M., Kääriäinen, J., & Teppola, S. (2017). Tackling the digitalization challenge: How to benefit from digitalization in practice. *International Journal of Information Systems and Project Management, 5*(1), 63–77.

Schwarz, N., Groves, R. M., & Schuman, H. (1999). Survey methods. In *Survey methodology program.* Institute for Social Research, University of Michigan.

Suau-Sanchez, P., Voltes-Dorta, A., & Cugueró-Escofet, N. (2020). An early assessment of the impact of COVID-19 on air transport: Just another crisis or the end of aviation as we know it? *Journal of Transport Geography, 86*, 102749.

Application of Blockchain Innovative Technology in Logistics and Supply Chain Management: A New Paradigm for Future Logistics

Almaz Sandybayev and Benjamin Silas Bvepfepfe

1 Introduction

Within the framework of the "Fourth stage of the digital revolution"—"Industry 4.0—Cyber-physical systems "—an important role belongs to digital logistics and Supply Chain Management (SCM). Moving to digital production and E-commerce makes to take a fresh look at logistics as a tool for managing value chains and to determine the focus of changes that must occur in logistics and SCM under the influence of the transition to cyber production.

If we bear in mind the changes already caused by IT technologies—changes in the structure of companies, the boundaries of companies/sectors/industries of the economy, a set of key competencies, business models, and business strategies, then Digital SCM/Logistics in these realities acquires strategic importance for combining business processes into a single infrastructure of the digital economy. From this perspective, Blockchain has been highlighted as vital important distributed secure technology in the twenty-first century. Blockchain technology is a distributed secure technology in the prevailing Industry 4.0 era of today and has attracted great attention from both academia and industry (Dobrovnik et al., 2018; Swan, 2017).

Blockchain is a multifunctional and multilevel information technology generally designed for reliable accounting for various assets and transactions. Blockchain is known as distributed ledger technology (Tschorsch and Scheuermann, 2016; Zyskind et al., 2015), which allows participants to secure the settlement of transactions, archive the transaction, and transfer assets at a low cost (Tschorsch and Scheuermann, 2016). Potentially, this technology covers everything without exclusion of the scope of economic activity and has many areas of change. These include

A. Sandybayev (✉) · B. S. Bvepfepfe
Faculty of Business, Higher Colleges of Technology, Abu Dhabi, UAE
e-mail: bbvepfepfe@hct.ac.ae

© The Author(s), under exclusive license to Springer Nature Switzerland AG 2022
J. Marx Gómez, L. O. Yesufu (eds.), *Sustainable Development Through Data Analytics and Innovation*, Progress in IS,
https://doi.org/10.1007/978-3-031-12527-0_6

finance and economics, operations with tangible and intangible other assets, accounting in government organizations and companies, logistics and supply chain management, and much more. However, there are still vulnerabilities and challenges related to this technology that should not be neglected (Berke, 2018).

To understand the prospects for the implementation of digital logistics / SCM, the paper researches the possibilities of blockchain technology and its applications focusing on the logistical part of the blockchain through a comprehensive literature review. The ultimate purpose of this research is to find out possible use of blockchain technology in logistics processes, to analyze impact of blockchain technology on business transparency and activity through scrutinizing a research question: why it is considered a timely transition to implementing blockchain technology in the supply chain? Businesses around the world are facing tremendous pressure related to COVID and post-COVID-19 implications, increased uncertainty, challenges, and constraints due to more unpredictable globalization, higher customer expectation, market competition, supply chain complexity, and uncertainty, which call for coordination and cooperation across the supply chains (inter- and intra-supply chains) and the needs for information technology (Huddiniah and Er, 2019).

This paper aims to highlight blockchain applications in supply chain management by researching the relationship between blockchain and supply chain collaboration and integration through observation of the Blockchain Enterprise Survey www. juniperresearch.com. Juniper Research is world-class recognized research, forecasting, and consultancy expert for digital technology markets. To answer the research question, a qualitative research is designed in this study mainly by adapting a desk research and reviewing the relevant literature, previous studies, and business cases in blockchain and supply chain.

2 Literature Review

First of all, it is necessary to analyze global trends in these areas of commercial activities and scientific research. Analysis of special literature and reviews of the largest consulting companies indicates what can be distinguished as the following main trends in the areas of knowledge:

1. *New digital technologies.* They cover the development of functionality in the field of global communication and information flows in supply chains. The most important innovation in this area is the ability to digitalize key business processes (including logistics), supported by sensor robotics and content information. Digitization allows organisations to accelerate an execution of business processes in supply chains, ensuring greater reliability and transparency of information for making informed decisions. This will lead to a cost reduction based on preventing possible risks and eliminating operations that do not add value to customers. The continuum of digital technologies that make up the main functionality of digital logistics includes Big Data (Big data processing and analytics), IoT (Internet of Things), technology blockchain (distributed transaction ledgers), Cloud services,

e-SCM, 3D Printing, and others. This trend is continuing with the emergence of new digital technologies, such as artificial intelligence (Davenport et al. 2020; Huang and Rust, 2020; Rai 2020), augmented and mixed reality (Hilken et al. 2020), Blockchain, Machine Learning, Internet of things (Hoffman and Novak 2018), robotics (Mende et al. 2019), and virtual reality (Sample et al. 2020), which are further transforming the way service firms develop and deliver their services to customers (Grewal et al. 2020a, 2020b).

2. *Analytics and modeling.* Analytics and modeling will be central to the future of SCM. Decisions will be based on real-time information rather than assumptions. Predictive analytics is now able to predict outreach moves accurately by using data along with appropriate predictive models. In the field of analytics and modeling, simulation model technologies are rationing, Big Data, OLAP, and in-memory will play an outstanding role, allowing the decision makers at all levels easily and quickly define scenarios and make optimal solutions. Big Data opens new perspectives that allow extracting value from Big Data. Big Data-based technologies are being applied with success in multiple scenarios (Mayer-Schönberger and Cukier 2014; Agrawal et al. 2011; McAfee and Brynjolfsson, 2012). Considering Nana, SAP technology ERP can be used to invoke In-Memory technology—the future of fast and reliable decision-making that already exists in large software companies.

3. *Supply chain segmentation.* Supply chain segmentation will replace today's "one fits to all" approach. Segmentation of the supply chain allows to really solve customer requirements, therefore customer orientation is not just a word—it becomes realistic. Several methods have been applied in an effort to solve the measurement and operationalization issues of supply chain segmentation approaches (Rezaei and Ortt 2013a, 2013b; Osiro et al. 2014; Akman 2015; Lo and Sudjatmika 2015; Rezaei, Wang, and Tavasszy 2015; Hudnurkar, Rathod, and Jakhar 2016; Ross, Kuzu, and Li 2016). Agile Supply Chain, low-cost supply chains, quality-oriented supply chains, right down to tailor-made services clients are the future of SCM. Leading companies such as Bayer (Chemicals) and BMW (Automotive) are now actively using their supply chain segmentation strategy. In terms of segmentation, there is a link with technological innovation: to implement a proper segmentation-oriented planning process, it is necessary to use modeling technology and create scenarios, allowing to determine the ideal setting of the supply chain for a segment. The second technological aspect in the field of supply chain segmentation can be found in the field of transactional processes. As mentioned above, for each segment real-time information is required to identify relevant events in the supply chain and to act on disruptions (technology SCEM—Supply Chain Event Management) to ensure consistent service levels (especially in high-speed market segments).

4. *Service orientation.* Service orientation is closely related to the segment trend described above but also due to the growing implementation of Service Centers in the SCM functionality. They take an important planning and controlling functions of supply chain (ideology of 4PL outsourcing). The goal is to ensure fulfillment of service level agreements agreed between contractors in the supply

chain (an example is LSA—a logistic service level agreement). Integration of global event information is the foundation for success in the supply chain in centralized structures—service centers. This allows the planning and controlling group of the SCM department to respond quickly to events in the supply chain based on online information and make the right decisions. To support this idea, companies in the early 1990s were increasingly becoming more customer-oriented through their supply chains (Deshpande et al., 1993; Shapiro, 1988; Spekman et al., 1998).

5. *Supply chain optimization.* Supply chain optimization is already widely supported by software tools (by such companies—system integrators as SAP, Oracle, IBM Infor). In particular, in the field of designing an optimal structure of supply chain, many optimization tools are available in the market. These tools are capable of applying scenario modeling and simulation modeling to determine the best possible supply chain. In addition, there is a large number of software products for optimizing inventory and production processes available on the market. Optimization area based on operations research methodology is the most advanced and mature area in CSM and logistics. In response, industry tries to implement better operational solutions to manage sustainable and risky free supply chains (Barbosa-Póvoa, 2012; Heckmann et al. 2015; Snyder et al. 2016; Lawrence et al. 2016).

6. *Improving the resilience of supply chains.* Improving the resilience of supply chains is a major trend on the agenda of company leaders, logistics directors, and managers of SCM. Researchers define supply chain resilience as a capability to react and adapt or withstand unforeseen incidents—a reactive approach (Pettit and Fiksel, 2010; Rice and Caniato, 2003; Williams, Ponder, and Autry, 2009). Pressure from constantly increasing price, the financial crisis and COVID-19 disruptions that we have experienced in recent years, dramatically reduced the resilience of many companies' supply chains. Currently understanding of the key sustainability business environment can also have a positive impact on the strategies to increase profits. Better return management processes, reverse logistics and recycling, focus on reducing energy consumption, and green logistics which reduce the burden on the environment and prevent the accumulation of waste—lead to significant cost savings. The efficient design of a resilient supply chain network structure, supported by planning and optimization enables planners to define more efficient, energy efficient, and resilient efficient supply chain structures. Optimization software for optimal structural design of supply chain networks supporting green logistics will increase the market share.

The above trends in the development of CSM and logistics are undoubtedly associated with the further digitalization of the domestic economy in general and SCM logistics in particular. The ultimate goal of digitalization of supply chains is to provide relevant and reliable information about the parameters of key business processes, violations (failures) during its implementation, potential problem areas caused by risks, as well as optimization of the parameters of the supply chain for the implementation of socio-economic tasks. Companies with a digital supply chain can better use resources, assets, people and inventory, and move them faster to where they are needed at any given time to keep costs down by actively responding to possible risks during the transportation and production of goods. Potential benefits of a fully realized digital supply chain include economic mission in every area—from resources, time, and money to reducing environmental load. Ideally, the digital supply chain should have processes, backed by appropriate technology, which monitor inventory levels in real time, interactions with counterparties, the location of goods and equipment failures, and also use this information to plan and execute operations with improved performance levels. Technologies such as GPS tracking, radio frequency identification (RFID), barcodes, smart tags, data location-based and wireless sensor networks play an important role in digital supply chain. In addition, cloud and blockchain technologies integrated with web services can unify information and processes to ensure traceability and transparency of the supply chain.

3 Methodology

A qualitative research was employed by addressing the research question and understanding the important attributes of the blockchain for supply chain and its future role. The interpretive approach and based on observing and analyzing data provided by the research and consultancy platform: Blockchain Enterprise Survey www.juniperresearch.com. The research is required to grasp the subjective meaning of phenomenon and social action (Bryman, 2012). Mixed research methods including desk research and case study were applied in the study. A desk research was conducted to identify the key concept and technique of blockchain regarding supply chain management.

4 Results

In this section, some results are presented providing a deeper look at blockchain technology including its applications in the supply chain integration and collaboration. Blockchain is multifunctional and multilevel information technology generally designed for reliable accounting of various assets and transactions. Potentially, this technology covers everything without exclusion of the scope of economic activity and has many areas of changes. These include finance and economics, operations with tangible and intangible assets, accounting in government organizations and companies, logistics and supply chain management, and much more. The changes made by the various parties are collected and stored in the database at regular intervals as bundled packets called "blocks" (Mattila and Seppälä 2017). There are many special applications for blockchain nowadays. The most famous is a bitcoin, the basis of cryptocurrency. At a minimum, blockchain is the most secure and not expensive alternative to any company. Swan (2017) argues that cybersecurity is one of the biggest drivers of blockchain adoption. In addition, the application of these technology can fundamentally change the interaction of companies in the chain deliveries, as customers will have direct information about products and services of counterparty companies. Investors will be able to finance companies without the need for financial markets; within companies, any information can flow without any friction and behavioral decisions and will be completely transparent. Transparency and visibility is another important attribute of blockchain (Frank et al., 2019). Business models will be radically changed and improved; cost structures will become more pro transparent and the execution of contracts is unimpeded.

The impact of blockchain technology is not limited to the corporate sector. In turn, the new technology today finds the greatest application in the state sector. Central banks are already using blockchain and the most innovative digital economies (Singapore, Estonia, and Sweden) are using blockchain to secure the relationship between the regulator and citizens. Mansfield-Devine (2017) asserts that blockchain provides a kind of chain of trust in the commercial world. Property registry, public contracts, enforcement, and public records will be automated and facilitated through blockchain in a very short time. Basically, blockchain is an innovative paradigm for coordination of any kind of activities, including for interorganizational coordination of counterparties of supply chains. Many experts believe that Blockchain is the biggest innovation that we observe today, comparable to the Internet or ubiquitous computerization (digitalization) of the economy. It is not only a new type of Internet infrastructure based on distributed applications but also a new type of supply chain network, which may provide a new paradigm for future business (Hackius and Petersen, 2017; Mansfield-Devine, 2017).

In technological terms (on the computational level) blockchain is the chain of "blocks"—a decentralized ledger of all transactions in the computer network. From the standpoint of the SCM, it allows to solve one of the most difficult problems in implementing interorganizational coordination: how to ensure the security (transparency) of the flow of information and the trust of counter supply chain agents. Any interaction of counterparties in the supply chain (transactions) related to the receipt and transfer of goods and information (orders) in the management of material and financial flows: transportation, warehousing and cargo handling, customs clearance, banking operations, payments, electronic commerce, contractual relationships, email, leasing, online auctions and the like, requires multistage control to ensure the accuracy and reliability of the transmitted information, as well as the compliance of goods and services with quality standards and conditions of contracts. Blockchain technology solves this problem in a simple way: a ledger that is a decentralized list of all transactions in the supply chains that are simultaneously shared by all members of the chain. Decentralized network can retrieve and validate the information through the network (El Yli-Huumo et al. 2016).

Figure 1 shows a simple diagram illustrating the application of the blockchain technology in the supply chain. According to leading analytical companies, blockchain will change globally the dozens of industries—from finance to cybersecurity and many technologies will be reinvented. Therefore, enthusiasts will strive to implement projects on blockchain in a variety of business areas, hence the favor of

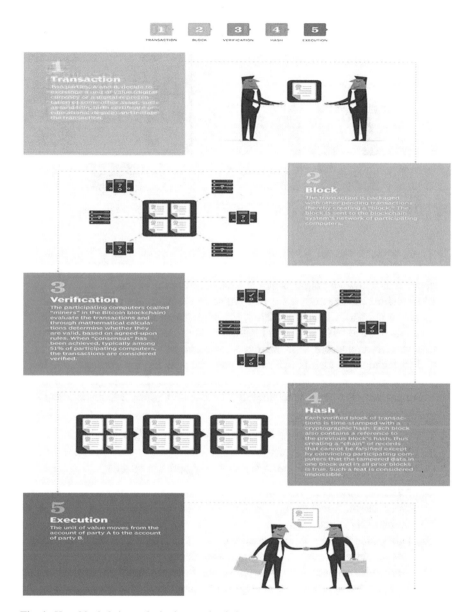

Fig. 1 How blockchain works in the supply chain

investors. The principle of decentralized distributed storage and transmission of data with a history of transactions that cannot be changed or destroyed, promises globally to transform the business, and given the possibility of implementing projects within the framework of global supply this will give quick feedback. Therefore, it is so important for entrepreneurs not to miss the moment and prepare for the new reality.

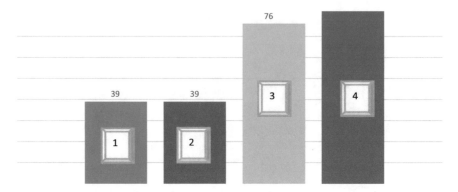

Fig. 2 Perception of the respondents (top management) of the Blockchain technology

However, some researcher, El Yli-Huumo et al. (2016) and Garrod (2016) argue that the future research directions for blockchain are not clear. It is possible to consider blockchain as a new paradigm which is a decentralized network.

Blockchain is fast becoming the main trend in the digitalization of the economy. However, as shown in a study for 2021, the implementation of blockchain technology increases up to 57% among companies with more than 20,000 employees. In particular, it is connected with raising awareness and understanding of what is related to technology distributed ledger, especially at the top management level. Figure 2 shows how the owners (founders), executive directors (CEO), and top management perceive the new technology "blockchain" (total 396 respondents):

1. Ready to deploy blockchain.
2. Top management is well aware of the capabilities of the blockchain.
3. Blockchain is useful/very useful.
4. Understand that blockchain provides several significant benefits.

   (a). Although 76% of all respondents believe that blockchain can be "useful" or "very useful," technology importance rose to 82% among founders of companies / executives.
   (b). Thirty-nine percent of all respondents said they are well versed in technology blockchain (47% among founders / managers).
   (c). Companies implementing blockchain solutions are rapidly advancing within the framework of per-rate strategies ("deployment" strategies). Among the companies that have achieved stages of PoC (Proof of Concept), two-thirds (66%) expected the blockchain to be integrated into their IT systems by the end of 2021.

Figure 3 shows the answers of 132 respondents (top managers of the largest public companies included in the Forbes 500). Almost a third of the respondents indicated that settlements and payments are the most popular cases of using blockchain technology, reflecting the status of the blockchain as a driver of financial instruments.

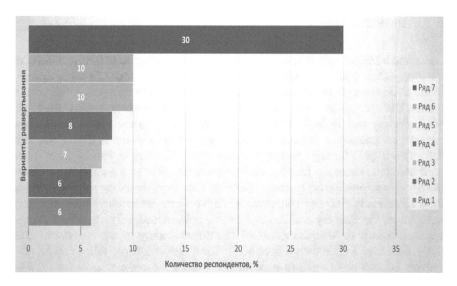

Fig. 3 Options dominating early deployment of blockchain technology. Source: Juniper Research

(1) healthcare; (2) data management; (3) identification; (4) Internet of Things (IoT); (5) supply chain management (SCM) and tracking; (6) Smart contracts; (7) settlements and payments

However, despite the fact that financial calculations and tools (in particular cryptocurrency/bitcoin) were one of the first use cases of blockchain, the universality of the technology is demonstrated by a number of applications identified by respondents. This refers to data management issues, "Internet of Things—IoT," Smart contracts, and so on. It was highlighted (10% of respondents) that logistics and SCM are rapidly evolving as a leading opportunity to deploy blockchain technology outside the financial sector.

In overall, blockchain is a digitalized, decentralized, public transaction book. What makes blocks different from other parameters is that every transaction is verified. An often cited property of blockchain like reaching "consensus" on transactions, eliminates the need for intermediaries and allows direct peer-to-peer transactions. This process is promising because it builds trust, minimizes the leakage of value into supply chain and creates a smoother playing field for competition.

Verified transactions are sent to an immutable ledger that is divided between enterprises/contractors in the supply chain. Each counterparty has a copy of the transaction log, so any change in any register will do it incompatible with others. This process makes it almost impossible to interfering into the transaction log (distributed ledger), since all participants can see all transactions. The blockchain thus provides transparency for all contractors in the supply chain.

Finally, smart contracts in blockchain technology provide governance by which the parties to this smart contract agree to interact with each other. Thus, smart contracts are the simplest form of decentralized automation in supply chains.

5   Practical Implication of Blockchain in Logistics and SCM

For example, in SCM the blockchain can completely revolutionize the functioning of the supply chain, radically simplifying and accelerating operations for counterparties of the chain throughout the network structure. Supply chain integration and collaboration has become an important trend in supply chain management (Chen et al., 2017; Soosay and Hyland, 2015). To illustrate the optimized information flow, consider an example of sending a batch of tulips from Holland to Singapore. This delivery will pass through several warehouses (hubs) using various types of transport. The route starts at a Dutch farm where a truck picks up the tulips and delivers them to the distribution center (hub). Since the counterparties of the supply chain can view all transactions in this shipment and delivery process without time-consuming negotiation, the delivery and carriage can be paid immediately after the ledger is updated to indicate that the shipment has arrived safely to the warehouse. The flowers are then loaded onto a liner bound for Singapore. Since tulips require cold storage, conditions of carriage may include fines if the temperature of the flower transport refrigerated trucks (which can be monitored by sensors) exceeds certain thresholds. If transportation conditions were violated en route, fines and penalties could be automatically levied, and insurance companies would be promptly notified of violations. Sensor readings available to all contractors in the supply chain, thus all parties to the process must be aware of condition of the cargo.

As it is seen from this example, the blockchain of transactions (blockchain) can track the flow and state of goods in the supply chain. Since every counterparty in the supply chain can view and trust the placed transactions, blockchain technology can reduce the time it takes to reconcile data and enable faster processes.

Another important example of the effective use of blockchain technology in logistics and SCM is a food security. According to the annual data of diseases recorded by the World Health Organization, one in ten people become ill worldwide and about 420,000 people die as a result of contaminated food. Retailers today are pushing for technology blockchain to provide customers with reliable food information, improve the traceability of their progress in the food supply chain, and increase consumer confidence in their stores.

To this end, IBM and the global retail chain Walmart pioneered the use of blockchain technology to develop a distributed ledger to transparently track food in supply chains. Blockchain allows users to track shipments of meat, perishables, and other products from suppliers to Walmart stores. Blockchain, alongside with a sensor system allows the retailer to track shipments quickly, efficiently, and reliably.

SCM is the integration of key business processes starting from the end user and covering all suppliers of goods and services and information that adds value to consumers and other stakeholders. Supply chain integration and collaboration have been identified as a key practice to achieve effectiveness and efficiency (Flynn et al., 2010; Wiengarten et al., 2016). The integration of these processes today needs adequate information support which can be provided through blockchain technology.

Today, even the largest organizations lack opportunities to resources and knowledge to deploy end-to-end information integration in their supply chains. Typically, the network structure of supply chains consists of many links (counterparties) which are permeated with material, information, and financial flows. All these flows can be managed in one unified information system, in which the participants in the supply chain have equal rights, since it is decentralized.

Speaking specifically about the industrial sector, the scope of blockchain technology is wide: from quality control to support of the entire product life cycle, ensuring the reliability and traceability of flow parameters for all participants in the supply chain. Blockchain application provides fast information integration between participants of supply chain and makes their relationship completely transparent. In modern business realities, information is usually not transferred along with the cargo, so it is difficult to make sure that many interested parties knew when the shipment would be delivered and plan operations in advance. This situation can be significantly improved using both electronic marking systems and systems based on a distributed ledger (Distributed Ledger Technology—DLT) or "blockchain." This technology permits to build distributed and multiuser real-time tracking, management of credit letters, and visibility of assets and liabilities. "Distributed book" uses blockchain technology, a shared digital ledger to continuously update the list of all transactions. Decentralized ledger keeps record of every transaction that takes place in the supply chain. With this technology, end users, among many other benefits will be able to trace their upload shipment in real time, view the stages of cargo movement on a single electronic card (Fig. 4).

One of the reasons for using blockchain technology in SCM is the fact that this technology can significantly reduce the share of counterfeit products on the markets. How it works? The entire life cycle of a product can be traced through a chain of transactions that are publicly available for of all participants in the supply chain, if it works on the "blockchain." If all industrial enterprises implement this technology, it will simply destroy all counterfeit products. Implementation of "blockchain" technology to a particular industrial company automatically protects it from counterfeiting—after all, these fakes will be impossible to register in the same distributed book mentioned above.

There are three key aspects of blockchain technology for SCM:

First, the shared accounts of verified transactions serve as the immutable single version of credibility (SVOT) for different companies. Ledgers such as SVOT have many uses, such as securing the origin of high-value goods, keeping a financial audit trail, and providing optimization solutions for SCM. One example is the recently announced partnership between IBM, Nestle, and Walmart. A group of large retailers and consumer goods manufacturers including Unilever, Nestle, and Walmart join IBM technology development blockchain project. Together with Unilever, Nestle, and Walmart, seven more companies joined the project: meat producer Tyson Foods, supermarket chain Kroger Co, manufacturer Dole Fruit and Vegetable, McCormick & Company Seasoning Manufacturer, Supplier Golden State Foods, Driscoll's fresh berry seller and cargo operator Carriage McLane Co. The partners plan to explore the possibilities of using blockchain to track the

Fig. 4 General view of the use of blockchain in the supply chain (Swan, 2017)

food supply chain to increase transparency and process safety. Also, the technology can be used to improve the reliability of data on the origin and condition of products when transferring information between by all participants in the supply chain, including manufacturers, suppliers, retailers, regulators, and consumers.

Using the blockchain allows quickly track the actions of several hundreds of participants in the process which will facilitate and speed up the detection of, for example, contaminated products and a potential source of contamination and then prevent the spread of goods that are dangerous to the consumer. Walmart is testing blockchain platform from IBM since October 2016. In June 2017, the retailer said that the technology has reduced the time it takes to track the movement of cargo with mango from 7 days to 2.2 s. According to Walmart representatives, one product review can cost a company from tens of thousands to millions of dollars in lost sales.

In fact, the widespread software solutions in SCM could be a "killer app" that actually promotes widespread blockchain adoption. However, it is important to note what needs to be addressed to the most important issue of data security and privacy management among participants in the supply chain before blockchain will be widely used in the SCM.

Table 1 Comparison of blockchain characteristics and supply chain structure

Lack of a central host		x
Immutable book		x
The only reliable version		
Systemic interaction		
Data security and privacy	x	
Strength, end-to-end supply chain	x	

Second, smart contracts on the blockchain enforce conditions and restrictions when transactions are carried out between participants. This process ensures consistency but more importantly, it ensures collaborative management and interorganizational coordination between supply chain partners.

Finally, blockchain shifts the focus of technological solutions from a separate company for the entire supply chain which serves as a system of interaction between contractors. This subtle point is by far the strongest impact on supply chains as the industry begins to realize that supply chains really are networks. SCM is a network problem requiring a network solution. The excitement around the blockchain raises awareness of supply chain counterparties and paves the way for faster implementation of operational solutions. Comparative characteristics of the blockchain (from the standpoint of the chain of transaction blocks) and the network structure of the supply chain is shown in Table 1.

As seen in Table 1, blockchain and supply chain networks provide SVOT and system of interaction between contractors. Whereas blockchain provides unique value to eliminate intermediaries and provide immutable book, supply chain network structures have already dealt with complex data validity issues and built a wide range of applications, covering key business processes, such as automatic control orders, VMI technologies, and integrated business planning.

Innovative applications outperform functional models of traditional point solutions such as order management, supply chain planning, transportation planning, transportation, warehouse management, and financial regulation. Moreover, the most advanced applications use multi-agent models and artificial intelligence in multiuser networks in real-time time is to achieve breakthrough performance. In fact, blockchain may simplify and automate the business transactions (Dobrovnik et al., 2018; Swan, 2017), e.g., reduce time and minimize risk in a complex supply chain.

6 Conclusion

The research discusses the possibility of using innovative information technology Blockchain in logistics and supply chain management. Today, Blockchain is a multifunctional and multilevel information technology designed for reliable accounting of various assets. Potentially, this technology covers all spheres of economic activity, without exception it has many areas of application. Among them finance, economic, and monetary settlements, operations with tangible and intangible assets. The implementation of Blockchain technology in any area of business implies the adoption of a completely transparent and reliable information platform that will be used by participants in a particular process: whether it is insurance of individuals, acts of real estate, management of returns, provision of logistics services, or any other types of activity. Within the framework of this chapter, the technology of "smart" assets and "smart" contracts is disclosed, examples of the use of cryptocurrencies in business are given. Until recently, it was difficult to imagine the real use of cryptocurrencies in any area of business, since this type of money is subject to high volatility and is associated with high risks. This would contribute to the further development of technology in the industry 4.0 era (Wang et al., 2020).

However, there are currently a number of different solutions to this problem. The key issue of the chapter is to uncover the possibilities of using Blockchain technology in logistics and supply chain management in general. Examples of the practical application of this technology in logistics are given, the advantages of Blockchain are revealed considering Blockchain Enterprise Surveys. Thus, the following important challenges are subject to be specified:

1. Both blockchain schemes and supply chain configuration optimization can disrupt the status quo and fundamentally transform the network structure of the supply chain.
2. Blockchain itself today does not have the power of advanced networked applications with layered data resolutions to optimally manage today's supply chains. However, it can significantly improve the transparency and security of the chain, thanks to its immutable ledger.
3. Used together, blockchain and digital business networks can provide secure multi-local applications that address the weaknesses of each technology while leveraging their strengths. Prolonging this situation, one can see natural convergence, where blockchain technology is inherent and essential part of SCM software applications.

References

Agrawal, D., Das, S., & El Abbadi, A. (2011). Big data and cloud computing: Current state and future opportunities. In *Proceedings of the 14th International Conference on Extending Database Technology, Uppsala, 21–24 March* (pp. 530–533). https://doi.org/10.1145/1951365.1951432

Akman, G. (2015). Evaluating suppliers to include green supplier development programs via fuzzy c-means and VIKOR methods. *Computers & Industrial Engineering, 86*, 69–82.

Barbosa-Póvoa, A. P. (2012). Progresses and challenges in process industry supply chains optimization. *Current Opinion in Chemical Engineering, 1*(4), 446–452.

Berke, A. (2018). How safe are Blockchains? it depends Harvard business review. Retrieved from https://hbr.org/2017/03/how-safe-are-blockchains-it-depends

Bryman, A. (2012). *Social research methods* (4th ed.). Oxford University Press.

Chen, L., Zhao, X., Tang, O., Price, L., Zhang, S., & Zhu, W. (2017). Supply chain collaboration for sustainability: A literature review and future research agenda. *International Journal of Production Economics, 194*, 73–87. https://doi.org/10.1016/j.ijpe.2017.04.005

Davenport, T. H., Guha, A., Grewal, D., & Bressgott, T. (2020). How artificial intelligence will change the future of marketing. *Journal of the Academy of Marketing Science, 48*(1), 24–42.

Deshpande, R., Farley, J., & Webster, F. (1993). Corporate culture, customer orientation, and innovativeness in Japanese firms: A quadrat analysis. *Journal of Marketing, 57*(2), 23–27.

Dobrovnik, M., Herold, D., Fürst, E., & Kummer, S. (2018). Blockchain for and in logistics: What to adopt and where to start. *Logistics, 2*(3). https://doi.org/10.3390/logistics2030018

El Yli-Huumo, J., Ko, D., Choi, S., Park, S., & Smolander, K. (2016). Where is current research on Blockchain technology? -a systematic review. *Plos One, 11*(10), e0163477.

Flynn, B., Huo, B., & Zhao, X. (2010). The impact of supply chain integration on performance: A contingency and configuration approach. *Journal of Operations Management, 28*(1), 58. https://doi.org/10.1016/j.jom.2009.06.001

Frank, A. G., Dalenogare, L. S., & Ayala, N. F. (2019). Industry 4.0 technologies: Implementation patterns in manufacturing companies. *International Journal of Production Economics, 210*, 15–26. https://doi.org/10.1016/j.ijpe.2019.01.004

Garrod, J. Z. (2016). The real world of the decentralized autonomous society. *Triple C: Communication, Capitalism & Critique, 14*(1), 62–77. https://doi.org/10.31269/triplec.v14i1.692

Grewal, D., Hulland, J., Kopalle, P. K., & Karahanna, E. (2020a). The future of technology and marketing: A multidisciplinary perspective. *Journal of the Academy of Marketing Science, 48*(1), 1–8.

Grewal, D., Noble, S. M., Roggeveen, A. L., & Nordfalt, J. (2020b). The future of in-store technology. *Journal of the Academy of Marketing Science, 48*(1), 96–113.

Hackius, N., & Petersen, M. (2017). Blockchain in logistics and supply chain: Trick or treat? In *Symposium conducted at the meeting of the Proceedings of the Hamburg International Conference of Logistics (HICL) Hamburg, Germany.*

Heckmann, I., Comes, T., & Nickel, S. (2015). A critical review on supply chain risk: Definition, measure and modeling. *Omega, 52*, 119–132.

Hilken, T., Keeling, D. I., de Ruyter, K., Mahr, D., & Chylinski, M. (2020). Seeing eye to eye: Social augmented reality and shared decision making in the marketplace. *Journal of the Academy of Marketing Science, 48*(2), 143–164.

Hoffman, D. L., & Novak, T. P. (2018). Consumer and object experience in the internet of things: An assemblage theory approach. *Journal of Consumer Research, 44*(6), 1178–1204.

Huang, M. H., & Rust, R. T. (2020). A strategic framework for artificial intelligence in marketing. *Journal of the Academy of Marketing Science, 49*(1), 30–50.

Huddiniah, E. R., & Er, M. (2019). Product variety, supply chain complexity and the needs for information technology: A framework based on literature review. Operations and supply chain management: An. *International Journal, 12*(4), 245–255. https://doi.org/10.31387/oscm0390247

Hudnurkar, M., Rathod, U., & Jakhar, S. K. (2016). Multi-criteria decision framework for supplier classification in collaborative supply chains. *International Journal of Productivity and Performance Management, 65*(5), 622–640.

Lawrence, V. S., Atan, Z., Peng, P., Rong, Y., Schmitt, A. J., & Sinsoysal, B. (2016). OR/MS models for supply chain disruptions: A review. *IIE Transactions, 48*(2), 89–109.

Lo, S. C., & Sudjatmika, F. V. (2015). Solving multi-criteria supplier segmentation based on the modified FAHP for supply chain management: A case study. *Soft Computing, 20*(12), 4981–4990.

Mansfield-Devine, S. (2017). Beyond Bitcoin: Using blockchain technology to provide assurance in the commercial world. *Computer Fraud and Security, 2017*(5), 14–18. https://doi.org/10.1016/S1361-3723(17)30042-8

Mattila, J., & Seppälä, T. (2017). Blockchains as a path to a network of systems - an emerging new trend of the digital platforms in industry and society. Retrieved from https://ideas.repec.org/p/rif/report/45.html

Mayer-Schönberger, V., & Cukier, K. (2014). *Big data: A revolution that will transform how we live, work, and think*. Houghton Mifflin Harcourt.

McAfee, A., & Brynjolfsson, E. (2012). *Big data: The management revolution*. Harvard Business Review.

Mende, M., Scott, M. L., van Doorn, J., Grewal, D., & Shanks, I. (2019). Service robots rising: How humanoid robots influence service experiences and food consumption. *Journal of Marketing Research, 56*(4), 535–556.

Osiro, L., Lima-Junior, F. R., & Carpinetti, L. C. R. (2014). A fuzzy logic approach to supplier evaluation for development. *International Journal of Production Economics, 153*(2), 95–112.

Pettit, T. J., & Fiksel, J. (2010). Ensuring supply chain resilience. *Journal of Business Logistics, 31*(1), 1–21.

Rai, A. (2020). Explainable AI: From black box to glass box. *Journal of the Academy of Marketing Science, 48*(1), 137–141.

Rezaei, J., & Ortt, R. (2013a). Multi-criteria supplier segmentation using a fuzzy preference relations based AHP. *European Journal of Operational Research, 225*(1), 75–84.

Rezaei, J., & Ortt, R. (2013b). Supplier Segmentation Using Fuzzy Logic. *Industrial Marketing Management, 42*(4), 507–517.

Rezaei, J., Wang, J., & Tavasszy, L. (2015). Linking supplier development to supplier segmentation using best worst method. *Expert Systems with Applications, 42*(23), 9152–9164.

Rice, J. B., & Caniato, F. (2003). Building a secure and resilient supply network. *Supply Chain Management Review, 7*(5), 22–30.

Ross, A. D., Kuzu, K., & Li, W. (2016). Exploring supplier performance risk and the Buyer's role using chance-constrained data envelopment analysis. *European Journal of Operational Research, 250*(3), 966–978.

Sample, K. L., Hagtvedt, H., & Brasel, S. A. (2020). Components of visual perception in marketing contexts: A conceptual framework and review. *Journal of the Academy of Marketing Science, 48*(3), 405–421.

Shapiro, B. P. (1988). What the hell is 'market orientation'? *Harvard Business Review, 66*(2), 19–25.

Snyder, L. V., Atan, Z., Peng, P., Rong, Y., Schmitt, A. J., & Sinsoysal, B. (2016). OR/MS models for supply chain disruptions: A review. *IIE Transactions, 48*(2), 89–109.

Soosay, C. A., & Hyland, P. (2015). A decade of supply chain collaboration and directions for future research. *Supply Chain Management: An International Journal, 20*(6), 613–630. https://doi.org/10.1108/SCM-06-2015-0217

Spekman, R., Kamauff, J., & Myhr, N. (1998). An empirical investigation into supply chain management: A perspective on partnerships. *Supply Chain Management, 3*(2), 53–67.

Swan, M. (2017). Anticipating the economic benefits of Blockchain. Technology innovation. *Management Review, 7*(10), 6–13. https://doi.org/10.22215/timreview/1109

Tschorsch, F., & Scheuermann, B. (2016). Bitcoin and beyond: A technical survey on decentralized digital currencies. *Communications Surveys and Tutorials, IEEE, 18*(3), 2084–2123. https://doi.org/10.1109/COMST.2016.2535718

Wang, M., Asian, S., Wood Lincoln, C., & Wang, B. (2020). Logistics innovation capability and its impacts on the supply chain risks in the industry 4.0 era. *Modern Supply Chain Research and Applications, 2*(1), 1–16. https://doi.org/10.1108/mscra-07-2019-0015

Wiengarten, F., Humphreys, P., Gimenez, C., & McIvor, R. (2016). Risk, risk management practices, and the success of supply chain integration. *International Journal of Production Economics, 171*, 361–370. https://doi.org/10.1016/j.ijpe.2015.03.020

Williams, Z., Ponder, N., & Autry, C. W. (2009). Supply chain security culture: Measure development and validation. *The International Journal of Logistics Management, 20*(2), 243–260.

Zyskind, G., Nathan, O., & Pentland, A. (2015). Decentralizing privacy: Using Blockchain to protect personal data. In *Symposium conducted at the meeting of the 2015 IEEE Security and Privacy Workshops*. https://doi.org/10.1109/SPW.2015.27

Industry 4.0: Drivers and Challenges in Developing the Smart Supply Chain Management

Nora Azima Noordin

1 Introduction

The rapid development in the industrial and manufacturing fields has revolved to the new stage that is known as Industry 4.0. The concept of Industry 4.0 has been initialized in the German strategic initiative in 2011 as part of its High-Tech Strategy 2020 aims. The main purpose of introducing the concept of Industry 4.0 is to build German manufacturing as an integrated industry lead market and provider. Based on the report from the Digital Transformation Monitor—Germany: Industrie 4.0 (Demetrius Klitou et al., 2017), Industrie 4.0 is a national level initiative from the German government through the Ministry of Education and Research (BMBF) and the Ministry for Economic Affairs and Energy (BMWI). Since digital manufacturing is currently in demand now, it aims to drive the concept of digital manufacturing to the different levels of operations including interconnection of products, value chains, and business models.

The Industrie 4.0 initiative is also aims with industry and academic collaboration in terms of networking of industry partners, standard and policy of Industry 4.0. The concept of Industry 4.0 has been introducing with the relations to the current evolution in manufacturing. Germany has set itself the goal of being an integrated industry lead market and provider by 2020 (Sharma, 2018). Nowadays, the manufacturing industries are changing from mass production to customize production. Industry 4.0 is a concept that everyone is familiar with before, but as the technology evolves, it becomes more advanced, complex, diversified, and at the same time able to connect manufacturing operations in real time. The capabilities realized by Industry 4.0 bring considerable benefits to companies including

N. A. Noordin (✉)
Faculty of Business, Higher Colleges of Technology, Sharjah Women's Campus, Sharjah, UAE
e-mail: nazima@hct.ac.ae

© The Author(s), under exclusive license to Springer Nature Switzerland AG 2022
J. Marx Gómez, L. O. Yesufu (eds.), *Sustainable Development Through Data Analytics and Innovation*, Progress in IS,
https://doi.org/10.1007/978-3-031-12527-0_7

customization of products, real-time data analysis, increased visibility, autonomous monitoring and control, dynamic product design and development, and enhanced productivity (Dalenogare et al., 2018).

The introduction of Industry 4.0 into manufacturing has a direct impact on the supply chain activities. Collaboration between suppliers, manufacturers, and customers are crucial to increase the transparency at all stages of the supply chain (Tjahjono et al., 2017). Supply chains can be defined as a network of facilities and distribution options that performs the functions of procurement of materials, transformation of the materials into intermediate and finished products, and the distribution of these finished products to customers (Ganeshan & Harriso, 2002). Supply Chain Management is now toward digitalization and the "Smart Supply Chain" is now being created with the adoption of advanced technology through Industry 4.0. Conventional supply chains consist of physical facilities scattered geographically to help establish and maintain transportation links among them. Thus, to transform from the conventional supply chain to the smart supply chain management, it requires a thorough assessment in terms of the drivers that are able to motivate the manufacturing firms and also the challenges that might encounter if the manufacturing firms consider to adopt the concept of Industry 4.0 in the supply chain management (Arnold et al., 2016). The concept of Industry 4.0 is to bring economic and social change in the world (Chauhan & Singh, 2019).

The purpose of this paper is to analyze the potential drivers and challenges in adopting the concept of Industry 4.0 with objective to create smart supply chain management using systematic review as the methodology (Tranfield, 2009).

2   Background

There were few stages of industrial evolution since 1800s where it began with the First Industrial Revolution, also known as Industry 1.0 in the late 1700s and early 1800s. It brought the transition from manual process to the first manufacturing processes and more optimized form of people labor through the use of water and steam-powered engines and other types of machine tools. In the early 20th century, the Second Industrial Revolution or Industry 2.0 revolution was taking place. The introduction of steel and use of electricity in factories was a new phenomenon at that time. The usage of electricity enables manufacturers to increase their efficiency and make factory machines more mobile. During this time, mass production concepts such as the assembly line were introduced as a way to increase productivity.

The Third Industrial Revolution, Industry 3.0 was slow to emerge as manufacturers started to have more knowledge, skills, and adoption of electronic chips and computer technologies in manufacturing systems. Manufacturing began to embed computer technology in their factories that makes factory machinery was operating under automation software and hardware. This was where digital technology started to grow.

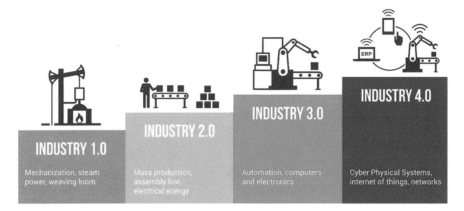

Fig. 1 The stages in industrial revolution

Today, the Fourth Industrial Revolution or Industry 4.0 has changed the atmosphere of how manufacturing operates as shown in Fig. 1. It is all in a digitalization environment with the assistance of the technology known as the Internet of Things (IoT), access to the real-time data, and the introduction of cyber-physical systems. IoT offers holistic approach to manufacturing operations by interlinked all the processes, from physical with digital and at the same time allow wider collaboration across department, vendors, suppliers, product and people.

2.1 The Nine Technological Pillars of Industry 4.0

The foundation of Industry 4.0 is based on the nine technological pillars that have been set up by the initiator. It consists of nine different elements that Industry 4.0 could help to transform the operational activities of the manufacturing sector as in description (Source: from website www.LCR4.uk).

2.1.1 The Internet of Things

The Internet of Things involves connecting the Internet to everyday items in order for them to send, receive, and process data. This can lead to a decrease in production time, aid risk management, and will save manufacturers valuable resources such as time and money.

2.1.2 Cloud Computing

This uses a network of remote servers to store, manage, and process data. This can be very beneficial to manufacturing businesses who can bring their own intelligence and knowledge to all sales situations as well as introduce faster new product development processes, releasing new products to market in shorter time periods.

2.1.3 Big Data

The use of vast volumes of data to improve efficiency and productivity is enabled by Big Data analytics. It helps organizations to gather value from large volumes of data to improve efficiency and performance of processes, increase flexibility and agility, and enhance product customization.

2.1.4 Cyber Security

Most manufacturers will want to protect their most valuable data including intellectual property, and data on customers and their products.

2.1.5 Systems Integration

This technology allows for different computer systems to be linked together, enabling actual communication, and the passing on of data between systems as software can act as a coordinated whole. This is ideal for all manufacturing companies as machinery from across the factory can be linked up together across the production line ensuring everything runs smoothly and efficiently.

2.1.6 Additive Manufacturing

Digital 3D design data is used to build a component in layers by depositing materials. Manufacturers may consider using 3D printing for lighter parts productions.

2.1.7 Autonomous Systems

Autonomous technology allows for machinery and robots, to act and behave autonomously after being programmed to do so. This technology allows for systems to think, act and react autonomously which also allows for decisions to be made remotely if control systems allow you to determine these behaviors from afar. This can all help contribute to a company's competitiveness, productivity, and of course, profitability.

2.1.8 Augmented Reality

Manufacturers could use this technology to showcase to customers what their products would look like without creating a physical copy. This way they are able to demonstrate what a product would look like or how it would work without the expense of a physical trial.

2.1.9 Simulation

This is a form of imitation of a situation, process, or environment. Many companies all over the world within a vast range of industries are starting to utilize virtual realities within their own businesses.

3 Analysis

3.1 The Design Principles in Industry 4.0

Industry 4.0 is the current trend of digital data exchange and automation through Cyber-Physical System (CPS) in manufacturing technologies (Savtschenko et al., 2017). It has principles that are set as a benchmark to support companies in identifying and implementing Industry 4.0 scenarios. Hermann et al. (2015) have mentioned six design principles that support companies in identifying possible Industry 4.0 pilots which can be implemented. The six design principles are Inter-operability, Virtualization, Decentralization, Real-Time Capability, Service Orientation, and Modularity, which will be discussed below.

3.1.1 Interoperability

It is the main enabler for Industry 4.0. In Industry 4.0 environment, CPS and humans are connected through Internet of Things (IoT). Standards will be the key to success for communication between CPS and manufacturers. The German government created the "German Standardization Roadmap" in 2013 to establish the standard use in CPS. Interoperability means that the CPS within the plant is able to communicate with each other using the recognized standard (Hermann et al., 2015).

3.1.2   Virtualization

CPS is able to monitor physical processes through virtualization. A virtual copy of the Smart Factory is created by linking sensor data with virtual plant models and simulation models. If there is any failure, a human can be notified. It also provides the next working steps and safety arrangements (Gorecky et al., 2014).

3.1.3   Decentralization

Embedded computers enable CPS to make a decision by itself. The RFID tags are able to send signal to the machines which working steps are necessary in Smart Factory. Therefore, central planning and controlling can be eliminated (Schlick et al., 2014).

3.1.4   Real-Time Capability

In the environment of rapid technology, it is vital for the organization to collect and analyze data in real time. The real-time capability provides immediate insights for data collection and analysis for decision-making purposes (Schlick et al., 2014).

3.1.5   Service Orientation

Service orientation offers services for Cyber-Physical Systems (CPS), humans and Smart Factories which available over the Internet of Services (IoS) that can be utilized by other participants. It can be offered both internally and across the company. As a result, the product-specific process operation can be composed based on the customer-specific requirements provided by the RFID tag (Schlick et al., 2014).

3.1.6   Modularity

Modularity offers a flexible adaptation of Smart Factories for the changing requirements of individual modules. Modular systems can be easily adjusted in case of seasonal fluctuations or changes in product characteristics. In the Smart Factory plant, new modules can be added using Plug & Play principles. Consequently, modular systems can be easily adjusted in case of seasonal fluctuations or if there are any changes in product characteristics (Hermann et al., 2015).

3.2 Industry 4.0 in Smart Supply Chain Management

Operations in supply chain management today has become more complicated and challenging due to the vast supply and demand around the world. Cheaper, faster, and better have always been the philosophy in managing supply chain activities (Wu et al., 2016). Based on the study conducted by Butner (2010), there are top five supply chain challenges that need to be addressed simultaneously. Firstly, the top challenges are cost containment. Supply chain executives rank cost containment as their number one responsibility to business. Secondly, is visibility as the greatest challenge in management? Next, is risk as supply chain executives agree on importance of risk management? Another challenge is in terms of customer intimacy. Rising customer demand ranks as the third-highest supply chain challenge and most of the companies are struggling to identify their customer needs. One of the top ranks challenges in supply chain management is globalization demand. Many companies are encountering issues in global sourcing as 65% in unreliable delivery. To deal effectively with the increasing challenges, the supply chain must become a lot smarter (Butner, 2010).

The element in Industry 4.0 should be implemented in the supply chain activities in a holistic. At the World Economic Forum Annual Meeting 2017 in Davos-Klosters, Switzerland, the Governors of the Supply Chain and Transport Community, and the Stewardship Board of the System Initiative on Shaping the Future of Production mandated the World Economic Forum (the Forum) to conduct consultations and research to understand the impact of the Fourth Industrial Revolution on the future of production and supply chains (Forum, 2017). Industry 4.0 enabled capabilities including highly organized interconnections and real-time monitoring and control of materials, equipment, and supply chain parameters that will help to improve the overall performance of the value chain and reduce risks (Ghadge & Moradlou, 2020).

This shows that the adoption of Industry 4.0 in the supply chain is important in order to overcome the challenges in managing the supply chain. According to (Wu et al., 2016), smart supply chain should possess the six distinctive elements as describe below and companies that embracing the elements of Industry 4.0 could be benefited from the smart supply chain operations.

3.2.1 Instrumented

Information in the next generation supply chain is overwhelmingly being machine generated, for example, by sensors, RFID tags, meters, and many others.

3.2.2   Interconnected

The entire supply chain including business entities and assets, IT systems, products, and other smart objects are all connected in a smart supply chain.

3.2.3   Intelligent

Smart supply chains make large-scale optimal decisions to optimize performance.

3.2.4   Automated

Smart supply chains must automate much of its process flows by using machines to replace other low-efficiency resources including labor.

3.2.5   Integrated

Supply chain process integration involves collaboration across supply chain stages, joint decision-making, common systems, and information sharing.

3.2.6   Innovative

Innovation is the development of new values through solutions that meet new requirements, inarticulate needs, or even existing needs in better ways.

3.3   Industry 4.0: Drivers and Challenges for Development of Smart Supply Chain Management

Many countries over the world realize the importance of embracing and adopting Industry 4.0 in manufacturing firms. This could result in improving the productivity and having a competitive advantage. Nations and manufacturing firms that lead in embracing Industry 4.0 technologies and processes will gain over global competitors. This competitiveness hinges on the ability to transform by responding to market shifts and technology trends (Malaysia Ministry of International Trade & Industry, 2018). In order for a manufacturing company to consider the adoption of Industry 4.0 in their manufacturing activities, specifically in the development of Smart Supply Chain management, there must be drivers that could motivate the companies to transform their operations. The manufacturing industry is increasingly shifting toward producing more technologically complex products. To compete with other rivals in industries are vital in order to sustain the business. Manufacturing companies must somehow take risks by adopting and embracing the new technology for competitive advantages. While companies are setting up for changes, there are some challenges that must be overcome and faced before they could fully adopt the Industry 4.0 technologies in supply chain management as a whole (Kiel et al., 2017). This paper has analyzed the potential drivers and challenges from the literature that could be considered by the manufacturing company in planning their transformation.

Tables 1 and 2 represent the analysis of drivers and challenges that manufacturing firms could consider in the process of adopting Industry 4.0 with the aim to create Smart Supply Chain Management. This paper has categorized the potential drivers and challenges based on the themes, drivers/challenges, and authors. Three main themes had been identified as supportive drivers in Industry 4.0, which are enabling ecosystem and efficient digital infrastructures; government regulation and policy and people or skill. From the analysis, it was found that there are seven main challenges that need to be addressed before the manufacturing firm could establish the Industry 4.0 in their supply chain operations. There are in terms of financial, people or skill, standard or requirement, culture or awareness or acceptance, digital technology readiness, government support, and organization and management.

4 Conclusion

This paper conducts a comprehensive review and presents the potential drivers and challenges in the adoption of Industry 4.0 in supply chain management. Industry 4.0 increased the digitization of manufacturing with Cyber-Physical System (CPS) in which connecting networks of humans and robots interact and work together with information shared and analyzed, supported by Big Data and cloud computing along entire industrial value chains.

In order to embark on Industry 4.0 environment, manufacturing firms must do an assessment and examine their potential and capabilities by identifying drivers and challenges that could motivate and hinder the transformation process. It will impact the whole manufacturing process and a new business model will be developed. Support from external and internal parties are vital to comprehend the practice especially in developing people skills, development or amendment on government regulation and policy, enable the ecosystem and digital readiness as well as increase awareness of need, benefits and opportunities of Industry 4.0 technologies and business processes among manufacturing firms. These could transform from the ordinary supply chain process to the Smart Supply Chain environment that will benefit the operations. If the vision of Industry 4.0 is to be realized, most business processes and systems must become fully digitalized. Many industry-leading companies have taken the step to implement advanced technology across their supply chains to improve visibility and bring down the walls between different business functions (The University of Warwick, Crimson, & Co, 2017).

Table 1 Industry 4.0: the drivers

Themes	Drivers	Authors
Enabling Ecosystem & Efficient Digital Infrastructure	• Fast and secure data connection is a basic requirement for the realization of Industry 4.0. A good and reliable Internet speed rate is needed for implementing Internet-based production technologies or services • IoT ecosystem and IoT Big Data are the most influential or driving IoT enablers. PCA, ISM, and DEMATEL are applied to model IoT enablers. • Supply Chain efficiency is boosted by the automation of physical tasks, planning and control, and information exchange process.	(Malaysia Ministry of International Trade & Industry, 2018) (Rajput & Singh, 2018) (Pereira & Romero, 2017)
Government Regulation and Policy	• Law in data integrity, security, and analysis are important areas of focus to ensure seamless data flow across value chains. • Financial support from Central Government is necessary for driving Industry 4.0. However, the main concern for firms is the level of security • Law and policy regarding employment: New labor and employment legislation is required to protect job losses and create a balance in the system.	(Malaysia Ministry of International Trade & Industry, 2018) (Bag et al., 2018)
People or Skill	• Technology users are critical success factors and play a decisive role in the implementation and diffusion of new technology. • Firms must start focusing on developing leaders with a new set of skills required in this digital world. Such management initiatives would lead to the better adoption of Industry 4.0 and drive sustainability in the supply chain network. • Collaboration with educational institutes: Companies have also shown interest in associating with educational institutes for Industry 4.0 technologies. • Enhance the capabilities of the existing workforce through national development programs specially designed for specific manufacturing sectors and support re-skilling and re-deployment. • Ensure the availability of future talent by equipping students with the necessary skillsets to work in the Industry 4.0 environment. • Strengthen the digital connectivity in and between industrial, educational, and training hubs to remove connectivity bottlenecks in adopting Industry 4.0 technologies.	(Müller et al., 2018) (Bauer et al., 2015; Gilchrist, 2016; Sung, 2018) (Aulbur et al., 2016) (Malaysia Ministry of International Trade & Industry, 2018)

Table 2 Industry 4.0: The Challenges

Themes	Challenges	Authors
Financial	• High investment levels • Unclear cost benefit	(Kamble et al., 2018), (Theorin et al., 2017)
	• High digital technologies investment for sensors, software, and applications • High employee training cost for organizational change purposes • Unclear economic benefit and digital investments	(Geissbauer et al., 2016)
	• Higher cost of adoption and longer payback period for Industry 4.0 technologies and processes	(Malaysia Ministry of International Trade & Industry, 2018)
	• Stagnating investments, expensive credit, and constrained access to global capital markets are some other financial obstacles apart from exchange rate volatility	(Aulbur et al., 2016)
People or Skill	• Lack of adequate skills required to handle the automation.	(Kamble et al., 2018), (Strom et al., 2016), (Aulbur et al., 2016)
	• Insufficient talent.	(Geissbauer et al., 2016), (Ghadge & Moradlou, 2020)
Standard or Requirement	• Lack of clarity in the standards for the implementation of Industry 4.0. • Lack of digital standards, norms, and certification. • It is essential for firms to focus on the development of a single set of common standards to support collaboration and reference architecture to provide a technical description of these standards. • Lack of proper documentation of the manufacturing processes followed at plants, hindering the efforts of automating processes. Even when the manufacturing processes have been documented, they vary from what is being actually followed on the shop floor.	(Kamble et al., 2018), (Strom et al., 2016) (Geissbauer et al., 2016), Branke et al., 2016) (Liao et al., 2017). (Aulbur et al., 2016)
Culture or Awareness or Acceptance	• Lack of digital culture and skills among employees in the organization. • Lack of awareness on the impact of and need for Industry 4.0 technologies. • Companies are well-advised to address employees' concerns and anxieties regarding data transparency, dependency on technical assistance systems, and workplace safety in human–machine interaction systems in order to enhance trust.	(Geissbauer et al., 2016), (Ras et al., 2017), (Schuh et al., 2017) (Malaysia Ministry of International Trade & Industry, 2018) (Müller et al., 2018)
Digital Technology Readiness	• Lack of integrated and digital approach to data gathering along manufacturing and supply chains. • Challenges in integrating supply chain	(Malaysia Ministry of International Trade & Industry, 2018)

(continued)

Table 2 (continued)

Themes	Challenges	Authors
	management and marketing from an information processing point of view.	(Ardito et al., 2018)
	• Challenges associated with the Internet of Things integration in Supply Chain.	(Haddud et al., 2017)
		(The University of Warwick, Crimson, & Co, 2017)
	• Challenges in how effective they are at developing an integrated supply chain approach that connects with suppliers and customers.	(Raban & Hauptman, 2018)
		(Bag et al., 2018)
	• To identify emerging technologies that are likely to have a significant impact on defense and attack capabilities in cyber security.	
	• Improved IT security and standards: The basic requirement for Industry 4.0 sustainability is security because the industrial control system is continuously interacting with the smart objects.	
Government Support	• Government guidelines and directions on Industry 4.0 in most of the economies are unclear. Governments are also unsure on probable consequences of Industry 4.0.	(Aulbur et al., 2016)
	• Currently, there is no central policy or a government body to drive Industry 4.0. Companies are taking individual initiatives like conducting seminars. Moreover, a comprehensive study evaluating pros and cons of Industry 4.0 has not yet been conducted.	
	• Government provides the necessary infrastructure for the digital world such as ICT infrastructure in most countries. However, there is lack of a roadmap for changing the industrial infrastructure, primarily due to lack of clarity and what are the benefits to the Industry 4.0.	
Organization or Management	• Unresolved questions around data security and data privacy in connection with the use of external data.	(Geissbauer et al., 2016)
	• Business partners are not able to collaborate around digital solutions.	(Kamble et al., 2018)
	• Slow expansion of basic infrastructure technologies.	(Malaysia Ministry of International Trade & Industry, 2018)
	• Concerns around loss of control over the company's intellectual property.	(Gokalp et al., 2017)
	• Lack of a clear digital operations vision and support/leadership from top management.	
	• The process changes which will be prompted due to the advent of CPS and smart factories.	
	• Organization functions may change	

(continued)

Table 2 (continued)

Themes	Challenges	Authors
	owing to automation Industry 4.0 will give rise to decentralized organizations. • Ownership of Intellectual Properties due to interconnectivity and information sharing along the supply chain. • Limited understanding of manufacturing firms of required future skills and expertise and own readiness to embark on Industry 4.0 transformation. • The Industry 4.0 transformational changes are quick and require adequate skill development and training which is challenging to deliver without high level of management support.	

References

Ardito, L., Petruzzelli, A. M., Panniello, U., & Garavelli, A. C. (2018). Towards Industry 4.0: Mapping digital technologies for supply chain management-marketing integration. *Business Process Management Journal., 25*, 323–346.

Arnold, C., Kiel, D., & Voigt, K.-I. (2016). How the industrial internet of things changes business models in different manufacturing industries? *International Journal of Innovation Management, 20*, 1640015.

Aulbur, W., Arvind, C. J., & Bigghe, R. (2016). Skill development for industry 4.0. In *Roland Berger, BRICS skill development working group* (pp. 1–50). India Section.

Bag, S., Telukdarie, A., Pretorius, J.-H., & Gupta, S. (2018). Industry 4.0 and supply chain sustainability: Framework and future research directions. *Benchmarking: An International Journal*, 1–42.

Bauer, W., Hämmerle, M., Schlund, S., & Vocke, C. (2015). Transforming to a hyper- connected society and economy – Towards an industry 4.0. *Procedia Manufacturing, 3*, 417–424.

Branke, J., Farid, S. S., & Shah, N. (2016). Industry 4.0: A vision for personalized medicine supply chain? *Cell and Gene Therapy Insight, 2*(2), 263–270.

Butner, K. (2010). The smarter supply chain of the future - insights from the global chief supply chain officer study. In *IBM Institute for Business Value and IBM Strategy & Change* (p. 38).

Chauhan, C., & Singh, A. (2019). A review of industry 4.0 in supply chain management studies. *Journal of Manufacturing Technology Management*. ahead-of-print.

Dalenogare, L. S., Benitez, G. B., Ayala, N. F., & Frank, A. G. (2018). The expected contribution of industry 4.0 technologies for industrial performance. *International Journal of Production Economics, 204*, 383–394.

Forum, W. E., (2017). Impact of the fourth industrial revolution on supply chains.

Ganeshan, R., & Harriso, T. P. (2002). *An introduction to supply chain management*. Springer.

Geissbauer, R., Vedso, J., & Schrauf, S. (2016). *2016 Global Industry 4.0 Survey Industry 4.0: Building the digital enterprise*. Pwc.Com..

Ghadge, A., & Moradlou, H. (2020). The impact of Industry 4.0 implementation on supply chains. *Journal of Manufacturing Technology Management, 31*(4), 669–686.

Gilchrist, A. (2016). *Industry 4.0 - The industrial internet of things*. Springer.

Gokalp, E., Sener, U., & Eren, P. E. (2017). Development of an assessment model for Industry 4.0: industry 4.0-MM. In *International Conference on Software Process Improvement and Capability Determination* (pp. 128–142). Springer.

Gorecky, D., Schmitt, M., Loskyll, M., & Zühlke, D. (2014). Human-machine-interaction in the industry 4.0 era. In *12th IEEE International Conference on Industrial Informatics (INDIN)* (pp. 289–294). *IEEE*.

Haddud, A., DeSouza, A., Khare, A., & Lee, H. (2017). Examining potential benefits and challenges associated with the internet of things integration in supply chains. *Journal of Manufacturing Technology Management, 28*(8), 1055–1085.

Hermann, M., Pentek, T., & Otto, B. (2015). Design principles for Industrie 4.0 scenarios: A Literature Review. *Working Paper, 01*, 16.

Kamble, S. S., Gunasekaran, A., & Sharma, R. (2018). Analysis of the driving and dependence power of barriers to adopt industry 4.0 in Indian manufacturing industry. *Computers in Industry, 101*(March), 107–119.

Kiel, D., Müller, J., Arnold, C., & Voigt, K. I. (2017). Sustainable industrial value creation: Benefits and challenges of Industry 4.0. In *ISPIM Innovation Symposium, the International Society for Professional Innovation Management (ISPIM). June, 1*.

Demetrius Klitou, Johannes Conrads, Morten Rasmussen, Laurent Probst, Bertrand Pedersen, P, & CARSA. (2017). Digital transformation monitor -Germany: Industrie 4.0.

Liao, Y., Deschamps, F., Loures, E. D. F. R., & Ramos, L. F. P. (2017). Past, present and future of Industry 4.0-a systematic literature review and research agenda proposal. *International Journal of Production Research, 55*(12), 3609–3629.

Malaysia Ministry of International Trade & Industry. (2018). *Draft National Industry 4.0 Policy Framework*.

Müller, J. M., Kiel, D., & Voigt, K. I. (2018). What drives the implementation of industry 4.0? The role of opportunities and challenges in the context of sustainability. *Sustainability (Switzerland), 10*(1), 247.

Pereira, A., & Romero, F. (2017). A review of the meanings and the implications of the industry 4.0 concept. *Procedia Manufacturing, 13*, 1206–1214.

Raban, Y., & Hauptman, A. (2018). Foresight of cyber security threat drivers and affecting technologies. *Foresight, 20*(4), 353–363.

Rajput, S., & Singh, S. P. (2018). Identifying Industry 4.0 IoT enablers by integrated PCA-ISM-DEMATEL approach. In *Management decision*.

Ras, E., Wild, F., Stahl, C., & Baudet, A. (2017). Bridging the skills gap of workers in industry 4.0 by human performance augmentation tools: Challenges and roadmap. In *Proceedings of the 10th International Conference on Pervasive Technologies Related to Assistive Environments* (pp. 428–432). ACM.

Savtschenko, M., Schulte, F., & Voß, S. (2017). IT governance for cyber-physical systems: The case of Industry 4.0. In *International conference of design, user experience, and usability* (pp. 667–676). Springer.

Schlick, J., Stephan, P., Loskyll, M., & Lappe, D. (2014). Industrie 4.0 in der praktischen Anwendung. In T. Bauernhansl, M. ten Hompel, & B. Vogel-Heuser (Eds.), *Industrie 4.0 in Produktion (Automatisierung und Logistik. Anwendung, Technolo- gien und Migration)* (pp. 57–84).

Schuh, G., Anderl, R., Gausemeier, J., Hompel, M., & Wahlster, W. (2017). Industrie 4.0 Maturity Index. In *Managing the Digital Transformation of Companies. Herbert Utz*.

Sharma, A.-M. (2018). *Industrie 4.0. Bundesministerium für Bildung und Forschung*. Retrieved from https://www.bmbf.de/de/zukunftsprojekt-industrie-4-0-848.html

Strom, M., Phillips, D., & Potter, J., (2016). *How digitization makes the supply chain more efficient, agile, and customer-focused*.

Sung, T. K. (2018). Industry 4.0: A Korea perspective. *Technological Forecasting and Social Change, 132*(October 2017), 40–45.

The University of Warwick, Crimson & Co, P. M, (2017). *An Industry 4 readiness assessment tool*.

Theorin, A., Bengtsson, K., Provost, J., Lieder, M., Johnsson, C., Lundholm, T., & Lennartson, B. (2017). An event-driven manufacturing information system architecture for Industry 4.0. *International Journal of Production Research, 55*(5), 1297–1311.

Tjahjono, B., Esplugues, C., Ares, E., & Pelaez, G. (2017). What does Industry 4.0 mean to Supply Chain? *Procedia Manufacturing, 13*, 1175–1182.

Tranfield, D. (2009). *Producing a Systematic Review*. The Sage.

Wu, L., Yue, X., Jin, A., & Yen, D. C. (2016). Smart supply chain management: A review and implications for future research. *International Journal of Logistics Management, 27*(2), 1–26. Retrieved January, 2020, from www.LCR4.uk

Reimaging Corporate Reporting in a Digital Economy Through Accounting Data Analytics

Kennedy Modugu

1 Introduction

The dynamic nature of the global economy remains apparently unflagging in its form and contents. The quest for businesses to create, deliver and sustain value in a stiffly competitive business environment has engendered a culture of inward-looking on the part of operators to develop modern smart business solutions that will not only reduce cost but deliver maximum convenience and satisfaction to customers and other stakeholders. The world is at the emerging phase of digital transformation that will completely obliterate the traditional way of doing things. Of the manifold and interesting challenges man faces at the moment, the most daunting and crucial is how to comprehend and design the new technological revolution which is the transformation of humankind. We are at the beginning of a revolution that is fundamentally changing the way we live, work, and relate with one another much more than what anyone has experienced in the past (Schwab, 2017).

No one can feign ignorance of the continuous changes that are driving the world of today and one of the fundamental drivers is digital transformation. At its core, digital transformation is not about Internet "unicorns." It is about using the latest technology to do what you already do, but better. The global economy is undergoing a digital transformation at a great speed (Deloitte, 2019). Therefore, the impact of these changes on existing processes and procedures should be a universal concern, especially to those professions that stand the risk of oblivion in the new global economy. As businesses get transformed through the "digital hurricane," so is the language of reporting. Accounting being the language of business responds to the ever-changing business ecosystem to meet the needs of stakeholders. The traditional

K. Modugu (✉)
Faculty of Business, Higher Colleges of Technology, Abu Dhabi, UAE
e-mail: kmodugu@hct.ac.ae

© The Author(s), under exclusive license to Springer Nature Switzerland AG 2022
J. Marx Gómez, L. O. Yesufu (eds.), *Sustainable Development Through Data Analytics and Innovation*, Progress in IS,
https://doi.org/10.1007/978-3-031-12527-0_8

accounting infrastructure can no longer handle the speed and volume of business data. Therefore, businesses must adapt to the new order to remain competitive.

2   What Is Corporate Reporting?

Annual reports and accounts of corporate organizations are the vehicles for communicating useful economic information about the operations of companies to diverse users. These users include shareholders, management, employees, suppliers, creditors, financial analysts, labor unions, stockbrokers, regulators, government agencies, and the public. These users rely on published financial and non-financial reports of corporate organizations to make informed decisions. Therefore, the quality of decisions made by users depends largely on the quality and quantity of available information. Corporate reporting is the process of recognition, measurement, and disclosure of corporate financial and non-financial information to permit informed judgment and decisions by users. The surge in the continuous demand for credible corporate reports by stakeholders is occasioned by the primacy of corporate reports as a principal tool for the communication of information to external users, assessment of economic performance, and condition of an enterprise in order to monitor management's actions and enhance the quality of decision-making.

The quality of corporate disclosure is fundamental to the efficiency of the capital market of any country and also critical to good corporate governance (Solomon, 2013). Improving the quality of accounting information is imperative to the health and vitality of financial markets, and a nation's economy at large (Al-Zarouni, 2008). Financial markets efficiently engender resource allocation if credible, reliable, and neutral financial information that fairly portrays the economic effect of transactions is available to participants in those markets (Herz, 2005). Okike et al. (2015) argue that the role and importance of disclosure and transparency in corporate accountability and governance cannot be overemphasized. Eisenhardt (1989) posits that while corporate disclosure has been the subject of much academic discourse, inadequate disclosure has been considered a key driver of agency complications in firms.

According to Biobele et al. (2013), to provide accounting information users with the required information needed for decision-making, the reporting jurisdiction needs to formulate quality standards and reporting practices to effectuate the process. Therefore, weaknesses in such standards and reporting practices hamper consistency, comparability, transparency, and a lack of trust in the information provided. This could lead to a high cost of capital and increased risks for different user groups (Al-Zarouni, 2008). Corporate transparency is determined by the information a firm discloses in the corporate report. Relevant and reliable disclosures are a means of corporate impression management and cost of capital reduction, thereby improving the marketability of shares. Compliance with mandatory corporate reporting requirements as well as the willingness to voluntarily disclose additional information above the statutory requirements signals the transparency posture of a reporting entity. In essence, a firm's level of disclosure affects the perceptions of the stakeholders about its operations. Accounting is often termed the language of business, and this language, like any other, must be communicated in accordance with the fundamental rules and in a manner that the recipient will understand.

Despite the collective efforts of the International Accounting Standards Board and the global financial community to improve the quality of corporate reporting which led to the introduction of IFRS and its subsequent adoption by a significant number of companies, the abuse of corporate reporting remains unabated across the globe. Evidence of the prevalence of serious accounting and reporting fraud includes BES 2014 scandal in which the giant Portuguese bank made a loss of €3.6 bn leading to the revocation of its operating license on account of serious misrepresentation and deliberate concealment of the bank's information by the directors. Also, the CEO of Toshiba, a Japanese company, Hisao Tanaka resigned alongside seven senior officers in 2015, after an independent inquiry revealed that the CEO had been aware that the company had inflated its profits by $1.2 billion for several years. This list goes on!

3 The Digital Economy

The digital economy is that which is created by the economic activities of a critical mass of everyday online connections among people, businesses, devices, data, and processes with little or no physical interference (Deloitte, 2019). It is a reversal of the traditional physical economy where a vast majority of processes are handled by humans. The bedrock of the digital economy is hyper-connectivity which means a growing interconnectedness of people, organizations, and machines that results from the Internet, mobile technology, and the Internet of Things (IoT). The digital economy is taking shape and undermining conventional notions about how businesses are structured, how firms interact, and how consumers obtain services, information, and goods.

This digital revolution has paved the way for a new era of information, sparking a fourth industrial revolution or "Industry 4.0" as it is also known (Schwab, 2017). It is mainly characterized by the processing of very large volumes of data with the aid of algorithms and mathematical models to support innovative technological solutions. This transformation is beginning to integrate business practices via "platform economy" and the emergence of global digital giants such as Google, Amazon, Facebook, Apple, Uber, Airbnb, Alibaba, and several others (Jeny, 2018).

Tom Goodwin, the senior vice president of strategy and innovation at *Havas Media* has this to say:

> Uber, the world's largest taxi company owns no vehicles. Facebook, the world's most popular media owner creates no content. Alibaba, the most valuable retailer has no inventory. Airbnb, the world's largest accommodation provider owns no real estate. . . Something interesting is happening (Goodman, 2015)

The First Industrial Revolution saw the use of water and steam power to mechanize production. The Second Industrial Revolution introduced the use of electric power to create mass production on a commercial scale. The Third used electronics and information technology to automate production. Now the Fourth Industrial Revolution is building on the Third digital revolution that has been occurring since the middle of the last century. It is characterized by a fusion of technologies that is blurring the lines between the physical-digital and biological spheres (Schwab, 2016). Unlike the successive industrial revolutions built on the ones before it, the Fourth Industrial Revolution is distinct in three fundamental ways: velocity, scope, and systems impact. The velocity of the current technological discoveries is unprecedented. Its global dominance and disruptive capabilities in almost every industry are remarkably intriguing. Of course, the breadth and depth of these changes signal the transformation of entire systems of production management and governance (Schwab, 2016).

Arguably, the birth of the Fourth Industrial Revolution and persistent digital transformations have shaped the corporate reporting ecosystem in terms of speed, content, and channel of communication. The traditional end-of-year financial reports are now available on a real-time basis with assets, liabilities, equity, income, and expenses captured with the latest digital corporate reporting tools. The advent of newer items of financial statement elements such as digital assets and their unique recognition, measurement, and disclosure added to the spectrum of corporate reports is a another evidence of the changing content and form of corporate reports. The goal of financial reporting is to provide decision-useful information to a vast array of users. For financial information to be useful, it must be both relevant and faithfully representative of the assertions of the management. Needless to say that for accounting information to be relevant, it must be comparable, understandable, verifiable, and timely.

With the introduction of accounting innovation to corporate reporting through accounting data analytics, the enhancing attributes of accounting information are speedily and better achieved, thus shortening the perennial problem of corporate reporting lag and making financial statements more transmissible, assessable, comparable, understandable, and verifiable using several accounting analytics tools.

4   Accounting Data Analytics

The traditional annual statement of profit or loss, statement of cash flow, statement of financial position, and statement of changes in equity can no longer match the speed of decision-making associated with the digital economy. Rather than wait for the comprehensive set of financial and non-financial statements for decision-making purposes, managers daily interrogate the gamut of big data generated at a very high velocity to chart the course of their businesses. In the same vein, other stakeholders now rely heavily on more readily available data to make their decisions. As the volume of accounting data continues its exponential trajectory, with its attendant burden of summarization and communication, business owners are devising means of bringing a large amount of information to stakeholders. Therefore, the need for data analytics infrastructure to capture, process, store, and communicate the information to users.

Data originates from different sources and in a variety of forms. Data from individual companies, organizations, streaming data, and so on accumulate in the cloud. Organizations access the information via in-house servers, laptops, tablets, and other mobile devices. Organizations may desire a unique combination of external and internal databases, including calculations, projections, and so on. That information is sent to multiple computers with individual processors to analyze the vast amounts of data.

5 What Is Big Data?

Big Data is a set of high-volume, high-velocity, and high-variety information that demands cost-effective, innovative forms of information processing for enhanced insight and decision-making. The end goal of Big Data should be to leverage the information, resulting in increased value to the customer and an organization. In the early twentieth century, businesses kept track of financial and operational results using paper and ink. It was difficult enough just to record the date of the transactions, let alone summarize information with financial statements (Lindell, 2017). The main form of automation that helped improve the efficiency of accounting clerks was limited to innovations in carbon copy paper, mimeograph machines, copy machines, and the like. When computers finally were available for operational and financial use, the systems were based on a batch recording of transactions. Again the focus was on capturing internal data to help an organization understand its financial and operational results. As computers advanced and became more powerful, the focus increased on obtaining more internally generated operational and financial information as well as analyzing the myriad information as a result of increased computing power, increased data, and more user-friendly tools.

Before the advent of the Internet, organizations worked mainly with their internal data. With the subsequent advances in Internet use in the latter half of the twentieth century and the beginning of the twenty-first century, external information became accessible that could be integrated with internal data. Companies moved from producing batch information to employees generating information (on both the corporate and the personal level), to sensors producing data about all aspects of our lives. This last point can be frightening because appliances, sensors, and different apparatuses are generating more data in shorter periods than ever before. This has resulted in a flood of information, the concept of Big Data, and predictive analytics.

6   The Accountant and Big Data

Before 1978, the responsibility of an accountant was dependent on the underlying accounting systems. The accountant recorded, processed, and analyzed transactions and reported on the financial results. Also, the accountant created additional value by evaluating results based on ratio and variance analysis and industry comparisons (Lindell, 2017).

In 1978, the world's first spreadsheet, VisiCalc, was created. As a result, accountants began to input transactional data into spreadsheets and created new analyses. Spreadsheets resulted in customized reports and an increase in the overall volume of data. Over the last three decades, spreadsheet usage has increased. Database applications and large enterprise systems have become more available, user-friendly, accessible, and powerful. The accountant who was data challenged before 1978 is today awash with so much data that has become overwhelming. Accountants are now drown in data, but majority of them they lack the knowledge that is hidden in the data stream.

Accountants tend to look at data from their traditional data perspectives of acquiring, gathering, categorizing, aggregating, analyzing, and reporting information. The availability of Big Data allows for increased complexity and ability to perform deep data analysis which was not possible previously (Fig. 1).

Lindell (2017) opined that when the history of Big Data is viewed through the lens of accounting, it can be categorized by the interaction of seven different areas:

- Bookkeeping
- Accounting
- Calculating machines
- Computers
- Internet

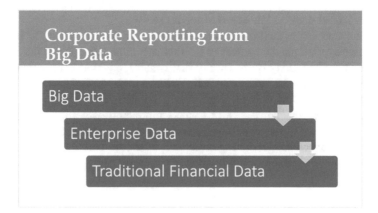

Fig. 1 From Big Data to financial statements

- Cloud computing
- Internet of Things

Each of these seven areas has been built upon or interacted with the other categories to result in the current access and application of Big Data. Consider some of the significant occurrences within each of the categories as they paved the way for Big Data. The first three categories remind us of where we came from. As you consider the fourth through fifth categories, itemize the impact that any of the items has made for you or your company.

6.1 Bookkeeping

Bookkeeping has been a part of human civilization from the very beginnings of recorded history. Chaldean-Babylonian civilization is attributed to having the first formal bookkeeping or recordkeeping activities. Archaeological evidence of the code of Hammurabi (leader of Babylonia from 2285 2242 B.C.) includes the following:

1. If a merchant gives an agent corn, wool, oil, or any other goods to transport, the agent shall give a receipt for the amount, and compensate the merchant. Then he/she shall obtain a receipt [from] the merchant for the money that he/she gives the merchant.
2. If the agent is careless and does not take a receipt for the money which he/she gave the merchant, he/she cannot consider the unreceipted money as his/her own.
3. If the agent accepts money from the merchant but has a quarrel with the merchant (denying the receipt), then shall the merchant swear before God and witnesses that he/she has given this money to the agent, and the agent shall pay him/her three times the sum.

In the earliest civilizations, transactions were recorded on clay tablets. During Egyptian times, transactions were recorded on papyrus. Systems continued to evolve through the Greek, Roman, and Israeli civilizations. Eventually, to promote the accountability of officials, public accounts were chiseled into stone. Records were used to facilitate transactions, tax assessments, and payments.

6.2 Accounting

In 1494, Fra Luca Pacioli (who wrote and taught in the fields of mathematics, theology, architecture, games, military strategy, and commerce) published *Summa de Arithmetica, Geometria, Proportioni et Proportionalita* (*The Collected Knowledge of Arithmetic, Geometry, Proportion and Proportionality*). A section of the book contained a treatise on bookkeeping, a section that ensured Pacioli's place in history as "The Father of Accounting," although he did not invent the practice of accounting, he described the process of double-entry accounting, known as the method of Venice. His system included most of the accounting cycle as we know it today. He published information on the use of journals and ledgers and described the process of debits and credits, including the fact that debits should equal credits.

As Pacioli says, this is the most important thing to note in Venetian bookkeeping: "All creditors must appear in the ledger at the right-hand side and all the debtors on the left" (Pacioli, 1494). All entries made in the ledger have to be double entries that is, if you make one creditor, you must make someone debtor." His ledger included assets, which comprises receivables, inventories, liabilities, capital, income, and expense accounts. The *Summa* was eventually translated into German, Russian, Dutch, and English.

Pacioli s system is the basis for the accounting systems still in use today. Fundamentally, not much has changed, even with the Industrial Revolution and the rise of corporations. However, a pivotal event in the world of Accounting started in 1913 with the ratification of America's Sixteenth Amendment. The amendment required a federal income tax to be paid by all individuals working in the United States. Income tax and corporate tax were little understood and heavily resisted in their formative years. As a result, most corporations and individuals were simply not filing or were filing incorrectly. A few years later in 1917, the Federal Reserve published *Uniform Accounting*, a document that attempted to set industry standards for how financials should be organized both for reporting tax and for financial statements. This was one of the earliest attempts at financial reporting standardization.

6.3 Calculating Machines

The history of calculating machines is nearly as old as the history of bookkeeping. The first calculating machine, the abacus, was in use as early as 2400 BC.

The early calculators of the seventeenth century laid the foundation for the computing revolution that would take place centuries later. Tabulating machines have been used by the accounting profession since the 1800s.

A few key dates from a photographic timeline on Gizmodo are outlined as follows:

- 1642: The Pascaline or Pascal s Calculator, by Blaise Pascal. It could add, subtract, multiply, and divide two numbers.
- 1820: The Arithmométre, the first mass-produced mechanical calculator, by Charles Xavier Thomas de Colmar.
- 1800s: The difference engines, the first mechanical computers, by Charles Babbage in the early 1800s and produced until 1972.

- 1948: The hand-cranked calculator Curta, invented by Curt Herzstark. Type II was introduced in 1954 and produced until 1972.
- 1971: Sharp Corporation invents the pocket calculator.

6.4 Computers

The advent of modern computers took place in the twentieth century. It revolutionized the accounting profession through increased speed, automation, and the introduction of spreadsheets. Although most of us witnessed much of the evolution first-hand, a timeline of key dates is outlined as follows:

1911: IBM is created.

1938: The Z1 computer was created, Konrad Zuse. It was a binary digital computer that used punch tape.

1947: William Shockley invents the transistor at Bell Labs.

1958: Advanced Research Projects Agency (ARPA) and NASA are formed. The first integrated circuit, or silicon chip, is produced by Jack Kilby and Robert Noyce.

1971: Ray Tomlinson invents email. Liquid crystal display (LCD) is developed by James Fergason. The floppy disk is created by David Noble with IBM. It is nicknamed the "floppy" for its flexibility.

1973: The Ethernet, a local area network (LAN) protocol, is developed by Robert Metcalfe and David Boggs. The minicomputer Xerox Alto was a landmark step in the development of personal computers.

1977: Apple Computer's Apple II, the first personal computer with color graphics, is demonstrated.

1980: IBM hires Paul Allen and Bill Gates to create an operating system for a new PC. They buy the rights to a simple operating system manufactured by Seattle Computer Products and use it as a template to develop DOS.

6.5 Internet

Considering how integral the Internet is to everyday life, it is hard to believe it is less than 50 years old. In 1969, the US Department of Defense set up the Advanced Research Projects Agency Network (ARPANET) with the intention of creating a computer network that could withstand any disaster. It became the first building block for what the Internet has become today.

In 1990, Tim Berners-Lee and Robert Cailliau proposed HTML hypertext proto-col for the Internet and World Wide Web. That same year, the first commercial Internet dial-up access provider came online. The next year, the World Wide Web was launched to the public.

A few more key dates in the development of the Internet:

1994: The World Wide Web Consortium is founded by Tim Berners-Lee to help with the development of common protocols for the evolution of the World Wide Web.

Yahoo! is created.

1995: Java is introduced.

Jeff Bezos launches Amazon.com. Pierre Omidyar begins eBay.

Jack Smith and Sabeer Bhatia create Hotmail. 1998: Sergey Brin and Larry Page begin Google. Peter Thiel and Max Levchin start PayPal.

Apple PowerBook G3 released. 2001: Bill Gates introduces the Xbox. Windows XP is launched.

2005: Blu-ray Discs are introduced. YouTube gets its start.

2009: Windows 7 released.

2012: Microsoft Windows 8 and Microsoft Surface are released.

6.6   Cloud Computing

Cloud computing is one of the newest developments in the use of the Internet. Instead of maintaining data and software on individual computers or local servers, data and programs can now be accessed in the "cloud." The term "cloud" was coined in 1997 to refer to the concept of shared data services and third-party access. Other recent developments in cloud computing are the formation of Amazon web services in 2002, Application Service Providers (ASPs) in 2005, and Hadoop in 2006.

6.7   Internet of Things (IoT)

The "Internet of Things" refers to the "network of Internet-connected objects able to collect and exchange data. Each object has a unique identifier and the ability to communicate machine-to-machine, without any human interaction. For instance, your home's intruder alert system automatically sends a signal when a lock is broken.

The following is a brief timeline of the IoT:

1832: Baron Schilling of Russia invents the electromagnetic telegraph. A year later, Germans Carl Friedrich Gauss and Wilhelm Weber invent a code to communicate over a distance of 1200 meters.

1961: GM introduces the first industrial robot, *Unimate,* in a New Jersey factory.

1969: ARPANET connects UCLA and Stanford universities.

1970: The Stanford Cart is unveiled, becoming the first smart car. Built for lunar exploration, it is controlled remotely and features a wireless video camera. The first handheld mobile cellphone goes on the market. It weighs 4.4. pounds.

1973: The first read-write radio frequency identification (RFID) tag is patented by Mario Cardullo. RFID tags will eventually lead to the wireless sensors so critical for enterprise, industrial, and manufacturing IoT technologies.

1993: The first webcam is created to monitor a coffee pot.

1994: Bluetooth is invented as an alternative to data cables to connect keyboards and phones to computers.

1999: Kevin Ashton, executive director of MIT s Auto-ID Center, coins the term "Internet of Things."

2000: Global Positioning System (GPS) becomes widely used by the public.

2001: Auto-ID Center proposes electronic product codes to identify every object in the world uniquely.

2005: Arduino simplifies interconnecting devices. 2006: Hadoop developed.

2010: Bluetooth low energy (BLE) is introduced, enabling applications in the fitness, health care, security, and home entertainment industries.

2011: Nest Labs introduces sensor-driven, Wi-Fi-enabled, self-learning, programmable thermostats and smoke detectors.

2011: Internet Protocol version 6 (IPv6) expands the number of objects that can connect to the Internet by introducing 340 undecillion IP addresses.

2014: Apple announces HealthKit and HomeKit, two health and home automation developments. The company s iBeacon advances context and geolocation services.

7   Categories and Characteristics of Data

Now that you are familiar with the history and rapid development of certain areas of technology, let us take a look at the seven categories we just learned about and the key characteristics of their data volume, information preparation, and the type of data analysis conducted. These categories and characteristics are especially interesting when accountants consider the implications for future accounting careers. A major component of accounting task was the preparation and massaging of numbers for decision-making purposes. Big Data analytics eliminates many of the lower-level accounting positions due to making the preparation and massaging of numbers unnecessary (Table 1).

8   Terminologies in Big Data Analytics

As in any new field, Big Data has some terms that must be mastered. The following list is not meant to be all inclusive, but it identifies many of the terms related to Big Data, analytics, and business intelligence.

Table 1 Evolution of record keeping to computing to today s big data

Description	Volume of data	How information is (was) created?	What type of data analysis?
Bookkeeping	Low/transaction	Manual	No time
Accounting	Low/transaction	Manual	Minimal time/manual reports
Calculating machines	Medium/transaction	Manual/small automation	Minimal time/manual reports
Computers	High/transaction/reporting	Manual/high automation	Volumes of automated structured analysis
Internet	High/transaction/reporting	Manual/high automation	Volumes of automated structured analysis + data detectives to find corroborating or competitive data
Cloud	High/transaction/reporting	Manual/high automation	Volumes of automated structured analysis + volumes of unstructured data + data detectives to find corroborating or competitive data + predictive tools to analyze volumes of data
Internet of Things	Unlimited/unstructured	Automatically generated sensors, and the like	Volumes of automated structured analysis + volumes of unstructured data + data detectives to find corroborating or competitive data + predictive tools to analyze volumes of data + automated applications to gather, archive, evaluate and predict data patterns, trends and strategic actions

Note. From Analytics and Big Data for Accountant (p. 2–12) by J. Lindell, 2017, Wiley. Copyright by AICPA

Business Intelligence (BI) The integration of data, technology, analytics, and human knowledge to optimize business decisions and ultimately drive an enterprise s success. BI programs usually combine an enterprise data warehouse and a BI platform or toolset to transform data into usable, actionable business information.

Data Analytics (DA) The science of examining raw data with the purpose of drawing conclusions from that information. Data analytics is used in many industries to allow companies and organizations to make better business decisions, and in the sciences to verify or disprove existing models or theories.

Cloud Computing A model for delivering information technology services in which resources are retrieved from the Internet through web-based tools and applications rather than a direct connection to a server. Data and software packages are stored in servers. However, cloud computing allows access to information as long as an electronic device has access to the web. This type of system allows employees to work remotely.

Dashboards A business intelligence dashboard (BI dashboard) is a BI software interface that provides preconfigured or customer-defined metrics, statistics, insights, and visualization into current data. It allows the end and power users of

BI software to view instant results into the live performance state of business or data analytics.

Data Mining The practice of searching through large amounts of computerized data to find useful patterns or trends.

Data Scientist An employee or BI consultant who excels at analyzing data, particularly large amounts of data, to help a business gain a competitive edge.

Data Visualization The presentation of data in a pictorial or graphic format.

Hadoop A free, java-based programming framework that supports the processing of large data sets in a distributed computing environment. It is part of the Apache project, sponsored by the Apache Software Foundation.

OLAP (OnLine Analytical Processing). OLAP is a powerful technology for data discovery, including capabilities for limitless report viewing, complex analytical calculations, and predictive "what-if" scenario planning.

Predictive Analytics The practice of extracting information from existing data sets to determine patterns and predict future outcomes and trends. Predictive analytics does not tell you what will happen in the future. It forecasts what might happen in the future with an acceptable level of reliability, and includes what-if scenarios and risk assessment.

Prescriptive Analytics A type of business analytics that focuses on finding the best course of action for a given situation and belongs to a portfolio of analytic capabilities that include descriptive and predictive analytics.

Semi-Structured Data Data that has not been organized into a specialized repository, such as a database, but that nevertheless has associated information, such as metadata, which makes it more amenable to processing than raw or unstructured data. For example, a Word document contains metadata or tagging that allows for keyword searches, but it does not have as much relational structure or utility as the information in a database.

Structured Data Data resides in a fixed field within a record or file. This includes data contained in relational databases and spreadsheets.

Unstructured Data Information that does not reside in a traditional row-column database. It often includes text and multimedia. Examples include email messages, word processing documents, videos, photos, audio files, presentations, web pages, and many other kinds of business documents. Although these files may have an internal structure, they are considered "unstructured" because the data is not contained in a database. Experts estimate that 80–90 percent of the data in any organization is unstructured.

9 Types of Data Analytics

Different types of analytics can be used to analyze Big Data for different purposes. They include the following.

9.1 Descriptive Analytics

Descriptive analytics is information that has happened in the past. From an accounting perspective, this would represent traditional historical financial information. Consider the following examples:

An assessment of customer credit risk can be predicted based on that company s past financial performance. A prediction of sales results can be created from customers' product preferences and sales cycles. Current product reviews can be used to predict future sales.

Employee evaluation can be used to predict turnover.

9.2 Diagnostic Analytics

Diagnostic analytics describes the reason for the historical results. It attempts to answer the question as in the following examples: In traditional finance, variance analysis uncovers the underlying reasons for differences in budgeted and actual results. Causal analysis can be used to describe why certain results occurred. Analytics dashboards can be used to describe why something happened. For example, with the COVID-19 pandemic, it was possible to view the daily spread of the virus as it occurred in different geographic regions. Tracking the increase in views, posts, fans, followers, and so forth, as a result of purchasing additional views on Facebook to increase the exposure of a particular post, video, or picture. Discovery analysis (insight) although not technically one of the four types of data analytics, could be inserted between diagnostic and predictive analytics. During discovery analysis or insight, research and analysis can be undertaken to identify whether there is a relationship between the historical information and another database.

9.3 Predictive Analytics

Predictive analytics attempts to determine what will happen by analyzing historical data and trends. Consider the following examples of predictive analytics: An accounting department prepares a cash flow projection report. Preparing an estimate of inventory levels Predicting an outcome based on changed assumptions. The revenue will increase by a specific percentage if an additional 5 percent is spent on the marketing budget. The issuance of additional coupons or promotions for a retail organization is projected to result in a 10% increase in revenue.

Based on historical results, ads released the week of Black Friday are predicted to generate greater than normal sales for the Black Friday holiday shopping season.

9.4 Prescriptive Analytics

Prescriptive analytics uses the information from descriptive, diagnostic, and predictive analytics to suggest specific decisions or changes in approach to a business strategy. It could also be described as the best scenario to take to achieve the desired outcome. The following are examples of prescriptive data analytics: Airline seat prices and the manner in which the cost per seat regularly increases as the departure date draws near. A related component of this decision is the airline's decision to overbook flights and offer incentives to placate passengers who are inconvenienced. Applications such as Facebook suggest to the user that there are additional friends they may wish to connect with. This "prescription for connecting" is based on the analysis of common friends in both of the individuals' profiles. Hence, the new friends and our potential friends are suggested as contacts. The most common prescriptive analytics would be medical drugs that have been known to alleviate certain medical issues (statin drugs, diabetic drugs, blood pressure drugs, and the like). The medications can also have negative predictions due to potential problem interactions.

10 Benefits of Big Data

Now that you know what Big Data is, you may be wondering how it will help you in your practice. What benefits can an organization derive from Big Data? A study from IBM26 showed that organizations competing on analytics outperform competitors by 1.6 times in revenue growth, 2.5 times in stock price appreciation, 2 times in EBITDA (earnings before interest, taxes, depreciation, and amortization) growth. Also, the World Economic Forum in 2012 stated that data gathering is a new class of economic asset, like currency and gold.

Other benefits of strategic benefits of Big Data to business include:

- Better strategic decisions.
- Quicker arrival of new products and services to market Increased innovation.
- Better insight into the business Better insight into the competition.
- Real-time change for existing products, services, or offers Environmental scans for threats or opportunities.

Big Data also aids in decision capability enhancement such as the following:

- Increase retained and analyzed the amount of data Increase the speed of data analysis.
- Produce more accurate results Better decision-making processes Improved forecasting.
- More accurate identification of root cause analysis.
- Smarter decisions leverage new sources of data to improve the quality of decision-making.
- Faster decisions enabled more real-time data capture and analysis to support decision-making at the point of impact, such as when a customer is navigating our website or on the telephone with a customer service representative.
- Decisions that make a difference focus Big Data efforts toward areas that provide true differentiation.
- Analysis based on entire data sets as opposed to sample sets.
- Enhanced transparency of data.

Businesses will also experience efficiency improvements, including the following:

- Reduce or eliminate manual processes Cost savings.
- Increased productivity.
- Automated routine decisions.
- Improved manufacturing productivity and maintenance.
- Integration of previously related databases Improved scalability.

Customer relationships and sales can also benefit from utilizing Big Data. Some of the benefits in these areas include the following:

- Improved customer satisfaction Better customer service Increased input from customers.
- Improved sales results via cross-selling and upselling Increased attraction and retention of customers Increased targeted marketing via social media.

Finally, Big Data can enhance a business' governance and compliance efforts through the following:

- Improved fraud detection.
- Improved risk assessment and management.
- Tools that can scan and access corporate data to prevent unauthorized release of data.

11   Sources of Big Data

There are a variety of Big Data sources that an organization can access. Consider Table 2 breaks down the sources of data among five different categories.

Table 2 Sources of data

Customer data Vendor data Product data Shipping data Accounting data Marketing data Employee data	Traffic count Attendees Number of calls RFID	Emails Surveys PDFs PowerPoints Documents Pictures Video Audio Consultant Info	Industry data Government data Almanac Benchmarking	Twitter LinkedIn Facebook Google search info Web RFID GPS Internet of Things Pictures Video Audio Edgar Industry data Government data Consultant info

Note. From Analytics and Big Data for Accountant (pp. 2–12) by J. Lindell, 2017, Wiley. Copyright by AICPA

Table 3 Reporting and processing of data

Known structured data	Known, unused structured data	Known unstructured data	Unknown structured data	Unknown unstructured data
Traditional IT systems	Identification	Analytics	Acquisition	Acquisition
Excel database reporting forecasting	Catalog curate relate predict	Identification catalog curate relate predict	Catalog curate relate predict	Analytics identification catalog curate relate predict

Note. From Analytics and Big Data for Accountant (p. 2–12) by J. Lindell, 2017, Wiley. Copyright by AICPA

From Table 2, the sources of data are broken down into five major categories:

1. Structured data that the company has.
2. Structured data that the company has but has not developed.
3. Unstructured data that the organization has.
4. Structured data that the organization does not have.
5. Unstructured data that the organization does not have.

Accountants are very familiar with the first and fourth categories. These categories are typically represented by traditional accounting and financial applications. Also, these categories are used to perform ratio and interpretive analysis to provide insight or understanding of structured data within these systems. Some companies may have explored the use of category two items (unused structured data that they have) especially if they have developed areas of emphasis such as key performance indicators.

The real challenge is for the organization to leverage categories three and five. (unstructured data.)

In Table 3, the tools required to access each of the categories are outlined as follows:

In category one, reporting and processing involve traditional accounting accumulation, reporting, forecasting, and analysis.

Category two requires the organization to identify what information is necessary to provide insight into the organization. Once identified, the company must catalog the information to allow access to it. The curator process prunes the necessary information and discards that which is not desired. Lastly, the information is related to other financial information to determine if it can be used to infer some impact on the financial performance of the organization.

Tools that can be used in the third category assess unstructured data that the organization has not considered before but currently has access to. Typically, the organization will run analyses to determine if the data are valuable for further analysis. If so, the organization follows the same pattern as category two.

In the fourth category, tools must identify structured data that is available outside of the organization but not collected. Once identified, it must be collected, transformed, or related to other existing data and analyzed for insights. An example of this would be obtaining industrial production statistics on a monthly basis and comparing that information with the monthly sales to determine if there is any correlation. The monthly industrial production statistics would also be made available for regular analysis of the organization activities (for example, sales).

The fifth category represents unstructured data that the organization does not have and that the organization does not know the value of. In this category, the organization must first acquire the unstructured data and, once acquired, evaluate with analytics to determine if there is any value in the data to predict the financial impact on the organization.

12 Conclusion

The velocity of digital transformation is at a spate more than what was ever experienced in human history, and this will continue to increase as people become more aware of better ways of getting things done. For business owners to remain competitive, they need to up their game to serve more sophisticated stakeholders in all forms or lose their market share to more innovative and proactive managers. The amount of data churned out of business activities both internally and externally has placed a new burden on accountants, thereby disrupting the orthodox financial reporting model and traditional financial reports.

Not only do stakeholders need more financial information, but they are also yearning for non-financial information in a very timely fashion as opposed to the yearly ritual of presenting year-end financial reports. The task before the twenty-first century accountants is the deployment of digital technologies to analyze the gamut of Big Data to meet the information needs of proprietary interests and those of other stakeholders. This is the essence of *Accounting Data Analytics.*

In this chapter, the author x-rayed how the digital economy, driven by digital transformation is raising the stakes for accountants and how the traditional corporate reporting model is longer tenable in the age of the Fourth Industrial Revolution. Therefore, Accounting Data Analytics is the panacea to synthesizing the humongous data to meet the relevance and representational faithfulness of financial reports.

References

Al-Zarouni, A. (2008). Corporate financial disclosure in emerging markets: The case of the UAE (Doctoral Dissertation, Griffith University, UK). Retrieved from https://www120.secure. griffith.edu.au/.../Al-Zarouni_2009_02Thesis

Biobele, S. B., Igbeng, I. E., & John, E. F. (2013). The significance of international corporate governance disclosure on financial reporting in Nigeria. *International Journal of Business and Management, 8*(8), 100–106.

Deloitte. (2019). What is digital economy? Unicorns transformation and the internet of things. Retrieved, from, https://www2.deloitte.com/mt/en/pages/technology/articles/mt-what-is-digital-economy.html

Eisenhardt, K. M. (1989). Agency theory: An assessment and review. *Academy of Management Review, 14*(1), 57–74.

Goodman, T. (2015). The battle is for the customer interface. Retrieved from, https://techcrunch.com/2015/03/03/in-the-age-of-disintermediation-the-battle-is-all-for-the-customer-interface/

Herz, R. (2005, December). *Remarks of Robert Herz- chairman, financial accounting standards board.* Presented to 2005 AICPA National Conference on Current SEC and PCAOB Reporting Developments.

Jeny, A. (2018). What challenges does the digital economy bring to accounting. Retrieved from https://arc.eaa-online.org/blog/what-challenges-does-digital-economy-bring-accounting

Lindell, J. (2017). *Analytics and big data for accountant.* Wiley. Copyright by AICPA.

Okike, E., Adegbite, E., Nakpodia, F., & Adegbite, S. (2015). A review of internal and external influences on corporate governance and financial accountability in Nigeria. *International Journal of Business Governance and Ethics, 10*(2), 65–185.

Pacioli, L. (1494). Summa de Arithmetica, Geometria, Proportioni et Proportionalita (Summary of Arithmetic, Geometry, Proportions and Proportionality). Paganinus de Paganinis. Retrieved September 2021, from https://library.si.edu/digital-library/book/summadearithmeti00paci

Schwab K. (2016). *The fourth industrial revolution: What it means how to respond.* Retrieved from, https://www.weforum.org/agenda/2016/01/the-fourth-industrial-revolution-what-it-means-and-how-to-respond

Schwab, K. (2017). *The, fourth, industrial, revolution.* Penguin, Books, Ltd.

Solomon, J. (2013). *Corporate governance and accountability.* John Wiley & Sons.

Responding to COVID-19 Emergency: The Packaging Machinery Sector in Italy

Lucio Poma, Haya Al Shawwa, and Ilaria Vesentini

1 Introduction

This study focuses on the packaging machines sub-sector, by far the most vital within the automatic instrumental machinery sector. It refers to a strategic sector of the Italian economy that involves the production of machines or plants that are often sold as entire production lines, and that may include over ten different machines for industrial production and packaging (including wrapping and realization of products), especially in the pharmaceutical and food sectors. The definition of packaging in this paper is not limited to the final packaging of products and for this reason it is part of mechatronics and instrumental mechanics.

The sector of packaging is characterized by a very strong propensity to export with a high competitiveness rate. Automatic packaging machinery went against the trend during COVID-19 in some subdivisions within the sector, such as food and pharmaceutical, while in other subdivisions, such as the automotive, a downward

This chapter is a revised and expanded version of a paper entitled *Responding to COVID-19 emergency: The packaging machinery sector in Italy* presented at the International Conference "Achieving Sustainable Development (ASD 2021): Through Business Innovations & Digital Technologies" on 10/19/2021.

L. Poma (✉)
Department of Economics and Management at the University of Ferrara, Ferrara, Italy
e-mail: lucio.poma@unife.it

H. Al Shawwa
Business Division at the Higher Colleges of Technology, Sharjah, United Arab Emirates
e-mail: halshawwa@hct.ac.ae

I. Vesentini
Manufacturing Economic Studies (MECS), Modena, Italy
e-mail: i.vesentini@mecs.org

trend was manifested. However, since the food and pharmaceutical sectors are the driving forces of the packaging sector in Italy, the sector as a whole withstood the impact of the pandemic well. The overall sector has generally performed well, yet, not all the sectors with the same weight.

COVID-19 has also led to rising demand for certain food and pharmaceutical goods, consequently linking this to a higher demand in the economy for automatic machines. In some cases, there has been an increase in demand that had to be satisfied simultaneously with the need to adapt to the new health restrictions imposed in the workplace (smart working and social distancing), which had an evident impact on production. In addition, there were logistical problems and some shortages in the supply of semifinished products and raw materials. This research seeks to understand how the companies studied in this paper responded to the COVID-19 emergency, their sentiments and expectations during the pandemic. This research also aims to understand whether Industry 4.0 enabling technologies has helped some of these companies in the process to follow new productive and innovative paths.

2   Literature Review

Industry 4.0 is a new industrial model that characterizes the Fourth Industrial Revolution. The term first appeared at Hannover Fair in 2011 when Professor Wolfgang Wahlster, Director and CEO of the German Research Center for Artificial Intelligence addressed the opening ceremony, and the term was proposed to develop the German economy (Mosconi, 2015). Since its origins, it has been used as a synonym for Cyber-Physical System (CPS) in the domain of production (Vogel-Heuser & Hess, 2016). The main features of CPS are based on the view to achieve dynamic necessities of production and to progress the efficiency and competence of the whole industrial sector. This system enables all the physical processes and information flows to be available when and where they are needed across holistic manufacturing supply chains, multiple industries, small and medium-sized enterprises (SMEs), and large companies (Zhong et al., 2017). Such systems can exchange heterogeneous data and knowledge leading to the application of solutions in the different levels of business processes. Such collaborating computational entities become the most effective lever to improve industrial performance. This takes the name of interoperability, the ability of two systems to understand each other using the same functionality. In such a situation, participating entities remain autonomous, so that any of them can easily be replaced by another with similar specifications without changing the functionality of the overall system (Chen et al., 2008). Industry 4.0 presents a unique challenge of efficiently transforming traditional manufacturing into smart and autonomous systems. This consist of integrating manufacturing systems, materials, machinery, operators, products, and consumers, advance interconnectivity and traceability across the whole product life cycle in order to safeguard the horizontal and vertical integration of networked Smart Manufacturing systems. The enterprise architecture consists of three sub-systems that interact with each other: (1) a physical sub-system that delivers products and services, including human and technical agents; (2) decision sub-system, that manages, controls, plans, and monitors; and (3) an information sub-system that supports the other two, and that processes, stores and retrieves data (Romero & Vernadat, 2016). Each of these sub-systems can itself be viewed as a complex system, so enterprise architecture can be configured as System of System (Ackoff, 1972; DiMario, 2010). Kotov (1997) defined "System of Systems as large scale concurrent and distributed systems that are comprised of complex systems."

Industry 4.0 is based on the creation of an integrative and collaborative environment. This advanced manufacturing model is represented by intelligent, virtual, and digital performance in large-scale industries and emerges as a disruption when compared with previous three industrial revolutions (Ortiz et al., 2020). Industry 4.0 can be defined as a new level of value chain organization and management across the product life cycle (Henning, 2013), or as a collective term for technologies and concepts of the value chain organization (Hermann et al., 2016).

There are numerous concepts and terms that relate to Industry 4.0. Some foundational concepts include Internet of Things (IoT), Industrial Internet of Things (IIoT), Big Data, Cloud computing, Smart Manufacturing, Additive Manufacturing Technologies (3D), Smart Factory, and Machine-to-machine (M2M). Internet of Things (IoT), which is the most widespread among Industry 4.0 manufacturing companies, connects humans with machines and integrates knowledge between organizations (Lu, 2017). In 1999, Kevin Ashton first used the term Internet of Things (Ashton, 2009), and defined IoT as exclusively recognizable consistent objects with radio-frequency identification (RFID) expertise, which have the ability to alter the world. Both computer systems and IoT need to process an impressive and immense quantity of structured and unstructured data. Therefore, IoT and Big Data are closely connected as the IoT environments work on the norm of recording data, then storing (Rialti et al., 2019), then computing, and retrieving it from the cloud whenever the user needs to access it (Russom, 2011). Thus, the huge volume of data generated by the IoT needs the involvement of big data in IoT in order to reveal patterns, trends, associations, and opportunities. The different flows of information can also lead to the creation of new products and services. Therefore, the so-called Big Data today is placed at the heart of many action plans of production and service companies. This immense amount of data is growing exponentially every year (Kaisler et al., 2013). Early definitions described Big Data as "the explosion in the quantity (and sometimes, quality) of available and potentially relevant data, largely the result of recent and unprecedented advancements in data recording and storage technology" (Diebold, 2003). De Mauro et al. (2015) proposed a shared formal definition where Big Data represents the Information assets characterized by such a High Volume, Velocity and Variety to require specific Technology and Analytical Methods for its transformation into Value. These three attributed characteristics of Big Data were: volume, velocity, and variety the so-called 3Vs of Big Data (Laney, & 3D data management: Controlling data volume, velocity and variety, 2001). In the following years, the definition was expanded to include additional Vs such as veracity (Bello-Orgaz et al., 2016) and value (Zikopoulos et al., 2013; Berman,

2013; Gantz & Reinsel, 2011). One of the main advantages of organizations and companies using big data is to create predictive models that can anticipate problems reaching to solutions that reduce future costs.

Cloud computing refers to the practice of using interconnected remote servers hosted on the Internet, offers especially to small and medium-sized enterprises (that do not have the resources to build their own Business Intelligence (BI)) with the possibility to make use of reliable resources of software, hardware, and infrastructure as a Service (IaaS) delivered over the Internet and remote data centers (Armbrust et al., 2010). Cloud computing is a powerful technology to perform complex and large-scale computing operations without the need to maintain expensive computing hardware, dedicated space, and software (Hashem et al., 2015), and for this reason it has been widely spread among organizations (Huan, 2013). Digital data is another widely used technology, as it represents a valuable resource for marketing pro-fessionals to direct advertising channels; for production companies to use machine learning for predictive maintenance and intelligent warehouse logistics; for large distribution enterprises to optimize the arrangement of products and traceability (Poma et al., 2020). Big data and Industry 4.0 interconnect in multiple areas facilitated by IoT. For example, combining the economies of scale of homogeneous and mass production with extreme product differentiation, providing products that satisfy a single customer, the so-called mass customization (Chen et al., 2009), which leads to a cost-efficient manufacturing system.

3 Methodology (Surveys 1 and 2)

The methodology used in this research included two phases. We made use of two surveys, both submitted to packaging companies one year apart. The first survey made at the beginning of 2020, administered by MECS to a highly significant sample of 135 packaging companies, representing 73% of the sector's turnover, aimed at detecting the situation and expectations of companies during the lockdown (Baraldi, 2020). The second, very recent, made in 2021, aimed at identifying how the possibilities offered by Industry 4.0 helped individual companies and within the supply chain in which it is integrated to react to the market. This included 30 com-panies (*representing more than 60% of the sector's turnover*). In particular, our attention has been paid to the possibilities offered by Industry 4.0 both within a single company and that of the supply chain in which it is integrated.

4 Context Analysis: The Automatic Machine Sector

The packaging sector is one of the most central sectors in Italy in terms of exports, internationalization, innovation, and use of Industry 4.0 enabling technologies. Italy is the first producer in the world in automatic machines that competes with Germany from year to year. The sector that consists of Italian manufacturers of packaging machines (wrapping and packaging) is a universe of 616 companies that employ 33,304 employees and generated a turnover of 8 billion euros in 2019, 79.0% of which was achieved on international markets. (6.4 billion euros). Within the Italian instrumental mechanics' sector, it ranks first in terms of exports and second in terms of turnover. Year 2019 was the fourth consecutive year of growth of +2.2% (+ 1.8% on the national level and + 2.3% abroad). However, in 2019, growth slowed significantly compared to the trend of the previous year (+ 9.4% in 2018). Yet, this is considered a positive result since in 2019 it was the only sector of Italian instrumental mechanics to grow. The first anticipations of the 2020 results led to an estimate of a 5% drop in turnover, which is still the smallest drop in the sector and considerably lower than the estimated recession for Italy (-8.9%).

In recent years, the number of active companies has decreased slightly (-2.4% in 2019), not due to business closures, but rather due to merger and acquisition processes. In fact, in the meantime, employment has continued to increase, rising by 2.1% in 2019 compared to 2018. In 7 years, the sector has created almost 7 thousand new jobs, from 26,348 in 2012 to over 33,300 in 2019. With regard to the industrial structure, companies with a turnover exceeding 25 million euros, which are 8.6% of the companies in the sector, generate 70.3% of the overall turnover of the sector. Small companies, with a turnover of up to five million euros, on the one hand make up 64% of the total number of firms, but on the other hand they only contribute 8.4% to the turnover of the sector. Company size also affects profitability per employee. The largest companies reach an average figure of 281 thousand euros and the smallest companies of 126 thousand euros. On average, the turnover per employee of the entire sector is 241 thousand euros. The "social capital" is also linked to the dimensional structure: in companies over 50 million euros in turnover, the share of graduates is almost double (and in particular engineers) compared to smaller companies. Thus, this has furthermore a decisive impact on the propensity to export. As for the export countries, the European Union remains the main export area (40% of the total), but with a progressively decreasing incidence over time, the smaller companies are more oriented toward the nearest European markets. In fact, the small businesses (up to 2.5 million euros in turnover) export 44% of their total volumes and 58% of this share goes to European Union countries. While for larger companies (over 50 million euros in turnover), which export 87% of their turnover, the EU share is only 36.3% (ISN-National statistics survey Ucima, 2021). Overall, the 24 largest companies in the sector (3.8% of companies) made 62% of the total Italian exports in the sector. The trend is reversed as the geographical distances lengthen and the Asian continent comes to account for 23% of exports among the largest and most structured companies, more than double the incidence it has in micro-companies (10.1%).

From a location point of view, most of the companies, and especially in terms of employment and turnover, reside in the Emilia-Romagna region, followed by Lombardy, Veneto, and Piedmont regions. Along the Via Emilia (the so-called packaging valley), 36% of industrial activities are concentrated, over 56% of employees and it

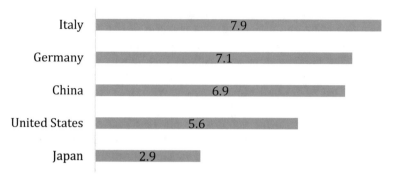

Fig. 1 Packaging machines: World Production (data in € billion). Source: Nomisma and MECS elaborations based on collected survey data

is home to more than 63% of the total turnover of the sector. Half of the "made in Italy" is made between Bologna, Parma, and Modena. Among the sectors that use packaging machines produced in Italy is the food sector that is in first place in terms of sales volumes, accounting for 29.6% of total turnover, followed by the beverage (26.4%) and the pharmaceutical and biomedical (18.3%). Cosmetics and chemicals occupied market shares of around 4%.

The world market for packaging machines has reached the threshold of 2020, strengthening an average growth of the CARG from 2015 to 2019 of 5.2% and a growth in turnover that has gone up in the last 5 years from 37 billion euros in 2015 to 45.5 billion in 2019. The top five producing countries make up 70 percent of world production and among these, Italy is in first place with a share of 18% equal to 7.9 billion euros (Fig. 1).

Over time, the ten-year uninterrupted growth in the sustained performance of this sector has made Italy become the leading country in the world in the production of automatic machines. As a result, the turnover of Italian companies more than doubled, from 3.8 billion euros in 2008 to 8 billion euros in 2019. The second country of production is Germany. Italy's fiercest competitor. These two countries hold leading positions through two diametrically opposed production systems. The Italian model uses a production organized by supply chain to create customized "tailor made" machines. The German model, which produces machines in series and tends to orient itself more toward the integrated enterprise.

In terms of numbers, the Italian companies are more than double the German ones, 631 compared to 250. But it is the distribution by size that reveals the clear difference between the two production systems. Eighty-two percent of Italian companies are small businesses compared to only 37 percent of German companies. On the other hand, large companies represent only 2 percent of Italian companies compared to 18% of large German companies (Fig. 2).

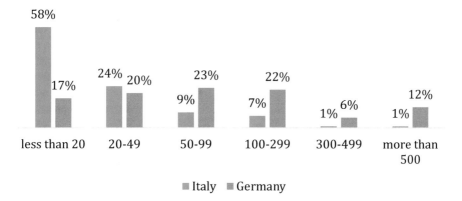

Fig. 2 Packaging machines: Dimensional comparison Italy versus Germany. Source: Nomisma and MECS elaborations based on collected survey data

5 Results

5.1 *How the Sector Suffered from COVID-19?*

5.1.1 Survey 1 Results

On the February 20, 2020, Italy recorded the first confirmed case of COVID-19. Two weeks later, on March 10th, the government decided to introduce a lockdown that only ended on May 18th. For 69 days, almost all economic activities were closed, as well as imposing stringent limits on freedom of movement. Essential and strategic activities were excluded from the production halt decree and were able to continue their production during the lockdown. These activities included the food and pharmaceutical sectors. The packaging sector according to the ATECO codes (used for the identification of essential and strategic sectors) pertained to the sector of automatic machines and was excluded from the derogation. However, the main packaging manufacturers supplied machines for the food and pharmaceutical sectors, and for this reason, they were included in the exemption and were able to continue their production. In this way, the sector, despite being included in the production halt decree, boasted a majority of companies that were able to continue their activities during the 2 months in which the country stopped production. The percentage of companies that have continued to operate was recorded at nearly 80%, a figure that we obtained from the two surveys carried out. The companies that have continued to operate have had to face two main difficulties. First of all, as a health precaution, workers had to be "spaced out" in the workplace and many of these were placed in smart working with negative repercussions on the production organization and on the productivity of the company. Secondly, there have been slowdowns and blockages in the supply of raw materials and semifinished products within the supply chain and along the value chain. Finally, some companies (for example, those linked to the pharmaceutical sector) have seen an increase in production, therefore, having to produce a greater quantity of machinery in an uncomfortable situation. Despite these dynamics, some companies experienced an increase in sales. It is necessary to understand how these companies reacted and forfeited during the lockdown in order to better interpret the actions taken by them to react in the first months of 2021.

The first survey made at the beginning of 2020 aimed at detecting the situation and expectations of companies during the lockdown, and was administered by MECS to a highly significant sample of 135 packaging companies representing 73% of the sector's turnover. During the COVID-19 emergency, 77.6% of the companies continued their production activities, 17.9% continued commercial or after-sales activities and only 4.5% stopped production. As anticipated, the most important Italian packaging chains produce machines for the pharmaceuticals, food, and beverage sectors which are extremely resilient, if not even growing sectors during the pandemic. Being connected to these sectors has partly safeguarded the Italian packaging sector.

In the first quarter of 2020, in the midst of the pandemic, just under 80% of the companies interviewed recorded a decline in turnover, production, and orders. However, there is 20% of companies that have not suffered any decline and even a small part has recorded an increase in turnover and production and above all in orders. As we have anticipated, the problems affected not only the demand side but also the supply side. Social distancing in companies, difficulty in the mobility of workers to the workplace, and the forced use of smart working have put companies in difficulty. 38.6% of the companies interviewed had a significant impact on productivity and for 49.6% had an impact, however, lower than the latter. Only 11.8% did not have an impact on productivity. For this last group, these companies have already partially implemented some technologies and "production environments" that are typical of Industry 4.0. For example, the companies that have an internal work unit dedicated to additive manufacturing (4.0). Given the high purchase cost, these large 3D printers are operational in a continuous cycle over 24 hours. At night, they are managed remotely, by the operator's house. In these cases, it was simply a matter of managing the systems from home throughout the day and not just in the evening. Others had already automated their systems through the CPS (Cyber Physical System) and their machinery was often linked with the intelligent warehouse whose semifinished products were selected and prepared by self-driving mobile units.

In all these cases (which unfortunately are not the majority), the companies have managed to react to the organizational difficulties of human resources in a much more effective way, such as not to reduce the productivity of their staff. The situation appears more worrying from the perspective of the supply chain and the value chain. Over two-thirds of the companies (81.8%) experienced slowdowns and problems with supplies along the supply chain and more than half of the companies (63.3%) had their own supplier closed. Health protection measures have hit the whole supply chain more than single companies. In Italy, the production chains are very complex, in which the various stages of processing are highly specialized and also entrusted to small or very small companies that have been most affected by the effects of the pandemic emergency. The reaction of the leading companies was timid, also slowed

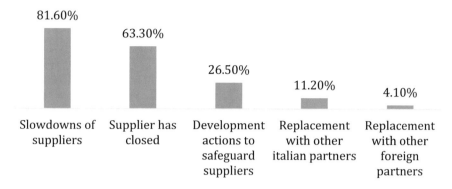

Fig. 3 Problems with the supply network. Source: Nomisma and MECS elaborations based on collected survey data

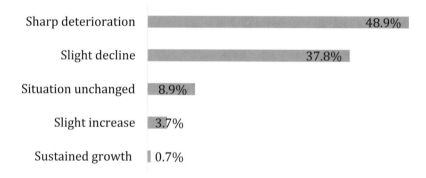

Fig. 4 Forecast for 2020. Source: Nomisma and MECS elaborations based on collected survey data

down by the general situation. A quarter of the companies have implemented actions to support the weaker companies in the supply chain. Fifteen percent replaced subcontractors: 11 percent Italians and only 4 percent with foreign subcontractors. This is because long segments of the Italian production chain use proximity economies, often aggregated in the form of small industrial districts. It is therefore very difficult and not immediate, to replace companies in the supply chain with foreign operators, the more specialized the processing phases become (Fig. 3).

But what were the sentiments of Italian Packaging companies in the first months of the pandemic? For almost half of the companies, the continuation of 2020 will cause a sharp worsening of the situation and for a third a slight decrease (86.7% in the total). But there is also 8.9% who expected an unchanged situation and, perhaps unexpectedly, there are also 3.7% of companies that anticipated a slight increase and 0.7% whose expectations are oriented toward a sustained growth of its production (Fig. 4). This last fact is central to our reflection. Companies that have adopted advanced Industry 4.0 technologies and new competitive strategies, even before

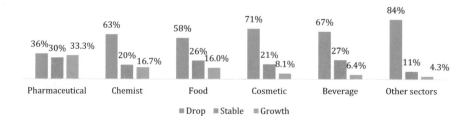

Fig. 5 Expectations for the different customer sectors. Source: Nomisma and MECS elaborations based on collected survey data

COVID-19, are the ones that have reacted best to the negative effects of COVID-19, both on the supply side and on the demand side.

As we have anticipated, the manufacturers of packaging machines are specialized in specific end markets, which also have an impact on the complexity and methods of the packaging machine produced. Some sectors are subject to stricter regulations than others, for example, they may need machines capable of carrying out more complex operations in a very short time.

In the pharmaceutical sector, certain segments during the pandemic experienced a growing demand for medicines. The expectations for 2020 were recorded absolutely positive (Fig. 5): a third of companies expected growth, a third stability, and a third a decline. This was followed by the chemical sector, partly linked to the latter, where 16.7 percent expected growth and 20% stability. Similar results were also recorded for the food sector (16%) that during the pandemic was resilient or growing, followed by cosmetics (8.1%) and beverages (6.4%). End markets, but also production specialization has affected how packaging companies reacted to the new situation.

5.2 How the Sector Reacted to COVID-19?

5.2.1 Survey 2 Results

After a disastrous year that recorded a recession of −8.9%, Italy has triggered a recovery path that shows more characteristics of growth rather than a simple rebound. Packaging (along with the pharmaceutical and food sectors) was one of the sectors that suffered the least from the harmful economic consequences of the COVID-19 emergency, even though some companies recorded equal or positive balance sheets in 2020. For this reason, the growth of this sector in 2021 assumes an even higher value of consolidation and being in an advantageous position. For this reason, we decided to carry out a second survey, which ended in September 2021 to understand the reason for this vivacity within the sector and if, and to what extent, the enabling technologies of Industry 4.0 have contributed to overcoming the emergency and to boost the sector on growth paths. Compared to the previous survey of 2020, the responding companies are smaller in number, 30 companies, however, the sample included all the large companies, including the leader of the sector, representing more than 60% of the sector's turnover.

5.2.2 Adoption of Industry 4.0 Technologies

The packaging sector is one of the most sectors, together with the automotive, embarking on the long transformation imposed by Industry 4.0. Additive manufacturing, augmented reality, 4.0 warehouses, and predictive maintenance of machines begin to spread. For example, in the pharmaceutical sector, where Italy boasts a production system in first place in Europe, the adoption of many 4.0 technologies is at a less advanced stage (Nomisma, 2020; Nomisma, 2021). Of the range of enabling technologies associated with Industry 4.0, Big data and artificial intelligence represent the sharpest point (those that require the largest investments) and are adopted by the largest size of the companies that use them (Fig. 6). They are also among those with greater potential, but whose yields, in terms of marginal gains (or turnover) mature later. Twenty-four percent of companies are already working with Big Data, and 33% are in the process of doing so, while less than half are not doing so (42%). The latter figure should not be read negatively, as in other sectors the percentages of companies that do not use big data are much higher. Similarly is

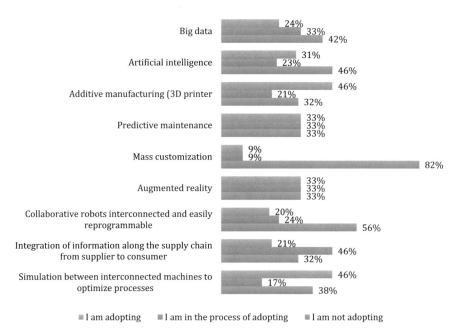

Fig. 6 Enabling technologies related to Industry 4.0 adopted by the companies. Source: Personal elaborations on MECS collected survey data

the situation for AI, where the companies that are adopting it are slightly higher than those adopting big data (31%), as well as those that are not adopting it (46%). It should be noted that few companies are using artificial intelligence algorithms to manage big data. AI is often present inside the machinery, which is often very complex and sophisticated and has to manage a full set of relevant data. Moving on to other enabling technologies, the most widespread is additive manufacturing (3D printer) used by almost half of the companies interviewed (46%). The reasons for such diffusion are manifold. First of all, it is a technology that has a very high-cost variability from 200 thousand euros for the smallest up to two million for the largest printer. Midsize businesses can approach this technology by purchasing a small, affordable printer first. Once we understand the advantages and potential in daily use, the company can strengthen itself by purchasing other printers of the same or larger size. Some companies have a real department dedicated to 3D printing which can also boast eight printers with a very high unit cost. As we will see later, 3D printing was a technology that, during the COVID-19 emergency, allowed a high-level of flexibility in production, never experienced before. For predictive maintenance, only a third of companies are embarking on this technology, which, in our opinion, will represent an important competitive turning point in the near future. From this study, it was found that all the companies already adopting big data and/or artificial intelligence technologies are also taking on the paths of predictive maintenance within their structures.

Predictive maintenance is a highly complex function that is being tested only on the most complex and expensive packaging machinery. It is a sophisticated algorithm, which processes all the data coming from IoT, plus the data of the different types of goods produced, the data of processing times and machine downtime, data related to maintenance, and many other climatic data (humidity and salinity of the air, etc.). It manages to anticipate the moment in which there would be the need to change a given element of the machinery before it breaks down causing the machine to stop.

Each day of machine downtime can be incredibly expensive for the company that has to pack the finished product. Since the largest companies in the sector usually sell an entire production packaging line (also consisting of ten machines in line), predictive maintenance, through the "dialogue" between the machines, extends to the entire line and not only to the single machine. It is therefore a question of systematizing the A.I. with a part of big data and for this reason, mainly large companies take this path. In relation to the production line, while almost half of the companies use the simulation between interconnected machines to optimize processes, only a fifth of the interviewees make use of interconnected and easily reprogrammable collaborative robots: the first technology is less expensive and complex than the second. Finally, this sector is still far from achieving mass customization: only 9 percent are adopting it, 9 percent plan to adopt it in the future and 82 percent are not adopting it. Mass customization aims to combine both the advantages of economy of scale and economy of scope that once were alternatives. It is the possibility of forming the circle between scale and scope using the whole range of possibilities offered by Industry 4.0 (Fig. 6). However, as we have previously

Fig. 7 Will the enabling technologies promoted by Industry 4.0 have a significant impact on the competitiveness of your company over the next 5 years? Source: Personal elaborations on MECS collected survey data

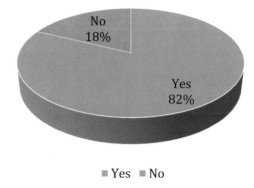

expressed, the strength of the Italian packaging sector lies in the ability to build customized machinery for the customer, which allows it to compete on an equal footing with the much more structured German sector in terms of average company size. Potentially, the Italian production system should be the one to benefit most from the future benefits of mass customization.

Industry 4.0 is not simply a bundle of enabling technologies but a revolution that insinuates itself into the deepest roots of the company. It becomes tough, if not impossible, to fully exploit the potential of these technologies if a deep reorganization of the company is not recognized. Changing your production organization, especially when it comes to a sector that has been winning on the market for many years, is not easy. It is necessary to overcome cultural obstacles, resistance to change, and investments in significant economic and human resources. However, it is an operation that must also be done considering that for 82% of the companies interviewed, the enabling technologies promoted by Industry 4.0 will significantly affect the competitiveness of their business over the next 5 years (Fig. 7). In this regard, we asked companies whether and to what extent the adoption of Industry 4.0 technologies has affected the organization of their business, also to understand whether this reorganization has helped the company to deal with the COVID-19 and post-COVID-19 emergency.

For 64% of companies, it was essential to create new organizational units, while only for 18%, Industry 4.0 was not very effective in this direction (Fig. 8). A common chorus on the need for new skills where only 6 percent consider this factor not very incisive. A waiting situation prevails over the possibility of creating new production lines, with the median placed on the value of 3 (45%) and the maximum and minimum equal to 10%. Such a balanced judgment between the parties shows that on the one hand there is great potential but on the other hand there are concerns about applications in a short time. If mass customization still appears to be a distant goal, the ability to customize the product is perceived as a tangible reality by 73% of the companies interviewed, while on the contrary, almost no company believes that Industry 4.0 will bring significant staff reductions. At this point, there is perhaps the biggest break between Industry 3.0 and Industry 4.0. While Industry 3.0 was mainly perceived by entrepreneurs as the automation revolution that would replace labor,

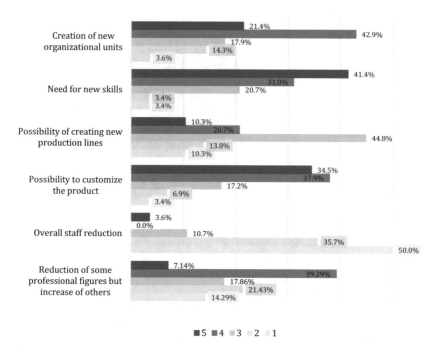

Fig. 8 How much do you expect Industry 4.0 to affect the following aspects of your internal organization? (grade from 1 to 5, with 1 = low incisiveness and 5 = maximum incisiveness). Source: Personal elaborations on MECS collected survey data

whose advantages were mainly in the reduction of labor costs, in Industry 4.0 it is perceived that the greatest advantage does not lie in a decrease in costs of production (which indeed they could increase) but in the infinite potential and flexibility of the system to be able to create new products in a new way. In fact, more than a decrease in personnel, entrepreneurs feel there will be a substitution effect of some professional figures with others.

5.2.3 The Reaction to the COVID Emergency and Smart Working

One of the most interesting aspects of this work was understanding whether, and to what extent, Industry 4.0 technologies contributed to addressing the COVID emergency. In general, for 76% of the companies interviewed, Industry 4.0 technologies were useful, while for 14% they did not help to cope with the problems resulting from the pandemic. The latter are mainly small companies and a few medium sized companies that are either currently not adopting Industry 4.0 technologies or in the process of adopting the technology. Within the companies that have benefited the largest share are companies that, thanks to these technologies, have optimized smart working in the best possible way. For example, companies that had already

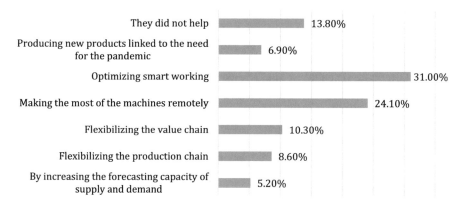

Fig. 9 During COVID, how did Industry 4.0 technologies help you to better face the situation? Source: Personal elaborations on MECS collected survey data

developed an automatic coding and employee pass system linked to Industry 4.0. Such a platform proved better management of the organization of work in the moment of health protection when employees had to shift and work respecting distances. In general, having used technologies related to Industry 4.0 to manage smart working was a significant advantage and of massive use, if we consider that 90% of companies made use of smart working during the emergency phase. For a third of the companies, smart working involved more than half of the staff and for another third from 25 to 50% of the entire workforce, while the last third involved 10–25% of the workforce. But the most relevant data is that 83% of companies did not use smart working before the health emergency. Therefore, a huge and extraordinary reorganization of work and processes had to be activated in a very short time and in this, the enabling technologies, for many companies, have proved to be a valid help and support. When we asked if these companies would like to continue using smart working, over a third have decided to continue using smart working even after the pandemic, despite never have used it previously.

For a significant share (24%), the advantage was to make the best use of the machines remotely (Fig. 9). Almost all additive manufacturing, given the high investment costs and the slowness in creating the programmed piece, is planned to work nonstop over 24 hours. Therefore, many of these were managed remotely already in the pre-COVID period, and the operator only intervened if the machine signaled a fault or jam, which could often also be remotely resolved. During the lockdown with reduced circulation of vehicles and people and with the social distancing of workers, these remote possibilities have allowed to continue that part of the process as if "nothing had happened." The incomplete pervasiveness of Industry 4.0 among companies does not yet allow for the potential flexibility within the supply chain and value chain to be exploited to the full. Only 10% of the companies interviewed benefited from the flexibility of the supply chain linked to new technologies and only 8% by extending the discussion to the value chain. The uses of big data and AI were still very scarce in the forecasting areas used to understand any shifts in demand by only 6 percent of companies. Very few (5%) but significant companies used new technologies to produce new products, which were totally diversified from their business, linked to the need to face the pandemic. Using the infinite potential of additive manufacturing, in one case, masks with a much greater filtering power than standard masks were produced. In another case, semifinished products suitable for the construction or maintenance of artificial respirators were produced, which was the scarcest good in hospitals during the acute phase of the pandemic.

When we asked if the pandemic has pushed the company to accelerate the introduction of technologies that can be summarized in the concept of Industry 4.0, the sample was almost split in half, 59% said yes and for 41% was a no. In addition, the pandemic has also had effects on the external context in which the company operates, in particular, the supply chain and the value chain. Among the main structural problems generated by COVID-19 in the supply network that stood out was the procurement of raw materials or semifinished products (38%), problems in logistics (31%), while 14% did not complain of particular problems. It should be noted that the pandemic has brought out problems that were already present, albeit sometimes in a latent way. When the value chain was put under stress, some links, which were already weak had finally collapsed. In fact, the problems highlighted remain with the same intensity even now that the health emergency is much less invasive and powerful. The reaction to these difficulties was immediate by mobilizing the internal forces of the productive system. Fifty-two percent of companies sought new suppliers in Italy, while 29% internalized part of the value chain by realizing some production phases of the supply chain on their own. Only 13% sought new suppliers across the border.

During the emergency phases, only 21% of companies have diversified their production activities but all of them believe they will consolidate this diversification also in the future. As in the case of technologies, a traumatic event has generated a push for renewal. Some companies have been pushed to diversify their products to seek new paths and new markets, understanding that even after an emergency, the initiated business could bring benefits even in a situation of returning to normal. In addition to diversifying production, companies have also acted by diversifying commercial channels. Forty-two percent of companies have activated web platforms or virtual guided tours for their customers, replacing the fairs in which they intended to participate. Seventeen percent invested in e-commerce while only 28% did not diversify their commercial channels. Finally, if on the one hand Industry 4.0 technologies have helped companies to cope with the health emergency, the pandemic has worked as an accelerator for the introduction of such technologies for 59% of the companies interviewed. There was a feedback mechanism. The new technologies have helped to cope with a traumatic change in the scenario which in turn prompted the introduction or greater use of the same.

6 Discussion and Conclusions

The analysis carried out leads to some reflections. First of all, Industry 4.0 technologies have helped packaging companies to cope with the COVID-19 emergency. This was particularly achieved through the remote management of machinery, and the interconnection between machinery also within the supply chain. This accelerated process was reflected in companies that were more ready on the front of Industry 4.0. Secondly, Industry 4.0 technologies, in particular additive manufacturing, have allowed some companies to recombine their knowledge by putting their production skills at the service of the country's pandemic emergency, in a completely different sector, for example, by producing masks with a particular technology or semifinished products essential for the construction of automatic breathing machines. Thirdly, the Italian packaging production system, organized by supply chain, has demonstrated extraordinary qualitative and quantitative flexibility, which, unlike the more rigid and integrated German production system, has allowed to better interpret the sudden changes in demand and provide supplies during the pandemic period. Finally, the COVID-19 pandemic has further incentivized companies to use Industry 4.0 technologies. Even when the emergency situation ended, the companies, which became more familiar with these technologies, declared their intention to continue investing in this direction. However, the analysis also shows that Industry 4.0 technologies are not yet adequately widespread among companies and an important difference exists between large and small companies in terms of their use of these technologies. Industry 4.0 plays a crucial role, especially in the packaging sector, which is a leading production sector in the country with high levels of technology, research and development, and is definitely export oriented. We trust that the huge European resources (195 billion euros) allocated to the Italian Recovery Plan can help Italian companies in this crucial technological transaction.

Acknowledgments We thank Chief Research Manager Luca Baraldi and Junior Analyst Giovanni Regano from Manufacturing Economic Studies (MECS), Modena, Italy for the administration and for the follow-up process of the two surveys directed toward packaging companies.

References

Ackoff, R. L. (1972). (1972), 'towards a system of systems concepts'. *Management Science.,* *17*(11), 661–671.

Armbrust, M., Fox, A., Griffith, R., Joseph, A., Katz, R., Konwinski, A., Lee, G., Patterson, D., Rabkin, A., Stoica, I., & Zaharia, M. (2010). A view of cloud computing. *Communications of the ACM, 53*(4), 50–58.

Ashton, K. (2009). That 'internet of things' thing. *RFID Journal, 22,* 97–114.

Baraldi, L. (2020). *Prospettive dell'Industria del Packaging.* MECS.

Bello-Orgaz, G., Jung, J. J., & Camacho, D. (2016). Social big data: Recent achievements and new challenges. *Inf. Fusion, 28,* 45–59.

Berman, J. J. (2013). *Principles of Big Data, DO,* 10.1016//C2012-0-01249-5.

Chen, D., Doumeingts, G., & Vernadat, F. (2008). Architectures for enterprise integration and interoperability: Past, present and future. *Computers in Industry., 59*(7), 647–659.

Chen, S. L., Wang, Y., & Tseng, M. M. (2009). Mass customization as a collaborative engineering effort. *Int J Collab Eng, 1*(1/2), 152–167.

De Mauro, A., Greco, M., & Grimaldi, M. (2015). What is big data? A consensual definition and a review of key research topics. In *AIP conference proceedings*. AIP.

Diebold, F. (2003). Big data dynamic factor models for macroeconomic measurement and forecasting. In *Advances in Economics and Econometrics: Theory and Applications, Eighth World Congress of the Econometric Society*. (edited by Dewatripont M, Hansen LP and Turnovsky S).

DiMario, M. J. (2010). *System of systems collaborative formation*. World Scientific Publishing Co. Pte. Ltd..

Gantz, J., & Reinsel, E. (2011). Extracting value from chaos. In *IDC's digital universe study*. EMC.

Hashem, I. A., Targio, I. Y., Anuar, N. B., Mokhtar, S., Gani, A., & Khan, S. U. (2015). The rise of "big data" on cloud computing: Review and open research issues. *Information Systems, 47*, 98–115.

Henning, K. (2013). *Recommendations for implementing the strategic initiative Industrie 4.0. Acatech*. National Academy of Science and Engineering.

Hermann, M., Pentek, T., & Otto, B. (2016). 'Design principles for Industrie 4.0 scenarios', *49th Hawaii international conference on system sciences (HICSS). IEEE, 2016*, 3928–3937.

Huan, L. (2013). Big data drives cloud adoption in enterprise. *IEEE Internet Computing, 17*(4), 68–71. https://doi.org/10.1109/MIC.2013.63

Kaisler, S., Armour, F., Espinosa, J. A., & Money, W. (2013). Big data: Issues and challenges moving forward. In *Proceedings 2013 46th Hawaii International Conference on System Sciences* (pp. 995–1004).

Kotov, V. (1997). Systems-of-systems as communicating structures. In *Tech. Rep. HPL-97-124, Hewlett Packard computer systems laboratory paper*.

Laney, D., et al. (2001). *META Group Research Note, 6*(70), 1.

Lu, Y. (2017). Industry 4.0: A survey on technologies, applications and open research issues. *Journal of Industrial Information Integration, 6*(2017), 1–10.

Mosconi, F. (2015). *The new European industrial policy: Global competitiveness and the manufacturing renaissance*. Routledge.

National Statistical Survey of Italian Packaging Machinery Manufacturers Ucima. (2021).

Nomisma. (2020). Osservatorio Nomisma sul settore dei farmaci generici anno 2020. Bologna.

Nomisma. (2021). Osservatorio Nomisma sul settore dei farmaci generici anno 2021. Bologna.

Ortiz, J. H., Marroquin, W. G., & Cifuentes, L. Z. (2020). *Industry 4.0: Current status and future trends*. IntechOpen.

Poma, L., Al Shawwa, H., & Maini, E. (2020). Industry 4.0 and big data: Role of government in the advancement of enterprises in Italy and UAE. *Int. J. Business Performance Management, 21*(3), 261–289.

Rialti, R., Marzi, G., Ciappei, C., & Busso, D. (2019). Big data and dynamic capabilities: A bibliometric analysis and systematic literature review. *Management Decision, 57*(8), 2052–2068.

Romero, D., & Vernadat, F. B. (2016). Enterprise information systems state of the art: Past, present and future trends. *Computers in Industry, 79*. https://doi.org/10.1016/j.compind.2016.03.001

Russom, P. (2011). Big data analytics. *TDWI Best Practices Report, Fourth Quarter, 19*(4), 1–34.

Vogel-Heuser, B., & Hess, D. (2016). Guest editorial Industry 4.0–prerequisites and visions. *IEEE Transactions on Automation Science and Engineering, 13*(2), 411–413.

Zhong, R. Y., Xu, X., Klotz, E., & Newman, S. T. (2017). Intelligent manufacturing in the context of industry 4.0: A review. *Engineering, 3*, 616–630.

Zikopoulos, P., Deroos, D., Parasuraman, K., Deutsch, T., Giles, J., & Corrigan, D. (2013). *Harness the power of big data: The Ibm big data platform*. McGraw-Hill.

Achieving Sustainable Development Through Green HRM: The Role of HR Analytics

Shatha M. Obeidat and Shahira O. Abdalla

1 Introduction

Based upon the Sustainable Development Goals (SDGs) that are set by the United Nation Development Program (UNDP), organizations serve as important stakeholders in achieving SDGs and are expected to utilize innovative technologies and systems for achieving sustainable results. Sustainable developments and particularly environmental sustainability have become an important agenda for today's organizations. Consequently, organizations are putting an increased emphasis on adopting green practices to promote sustainable performance.

The concept of Green Human Resource Management (Green HRM) has emerged as part of green practices implemented by organizations today. Green HRM includes HR practices that are aligned with environmental management practices in a way that could help improve organization's sustainable performance (El-Kassar & Singh, 2019). Nowadays, organizations invest heavily in human capital in an attempt to achieve sustainable performance (Boudreau & Cascio, 2017). Substantial work on the link between Green HRM and organizational performance took place and empirical evidences have shown that Green HRM is related to different sustainable organizational outcome measures (Obeidat et al., 2020). Accordingly, organizations today are looking for better ways to improve their HR system and processes.

S. M. Obeidat
Department of Management and Marketing, Faculty of Business, University of Qatar, Doha, Qatar
e-mail: Sobeidat@qu.edu.qa

S. O. Abdalla (✉)
Department of Human Resource Management, Faculty of Business, Higher Colleges of Technology, Abu Dhabi, UAE
e-mail: sosama@hct.ac.ae

© The Author(s), under exclusive license to Springer Nature Switzerland AG 2022
J. Marx Gómez, L. O. Yesufu (eds.), *Sustainable Development Through Data Analytics and Innovation*, Progress in IS,
https://doi.org/10.1007/978-3-031-12527-0_10

The role of HR analytics in achieving HRM goals cannot be neglected. The increasing use of data-driven HRM and advanced analytics systems has enabled organizations to integrate HR performance with business value and organizational performance (Margherita, 2021). In particular, the extensive use of objective facts and logical analysis rather than subjective evaluations can improve decision-making in HRM. Applying this on Green HRM, HR analytics can adopt advanced data analysis and visualization models and techniques to enhance strategic decisions related to Green HRM issues, thus serving the needs of top management to achieve sustainable performance.

In this book chapter, we will explore the contribution of Green HRM in achieving green performance and promote circular economy. To do so, we will begin by exploring the environmental dimension of sustainable development. Then, we will discuss how Green HRM functions and practices can help improve sustainable performance and how it links to circular economy system. Moreover, the role of HR analytics in enhancing strategic decisions about several green HR functions will be discussed later in this chapter.

1.1   The Environment Dimension of Sustainable Development

The environment dimension is part and parcel of sustainable development, originating from the French term *soutenir*, "to hold up or support" (Brown, 1991). The popularity of sustainable development, including the environment, emerged in 1983, following the creation of the International Commission on Environment and Development in accordance with the UN General Assembly's resolution 38/161 (United Nations, 1992). The RIO +20 Summit in 2012 resulted in the transition from Millennium Development Goals to Sustainable Development Goals (SDGs). The United Nations SDGs include development goals aimed at achieving sustainable management, efficient use of natural resources, mitigation of pollution and its impact, implementation of practices that prevent waste generation, and prioritizing recycling processes (United Nations, 2015).

Fischer-Kowalski and Swilling (2011) claimed that global sustainability occurs through detachment of economic growth rates and resource consumption, to enhance productivity through technology, social, and organizations innovations. We support Zhykharieva et al. (2021) arguments that environmental resilience cannot be achieved without organizations' commitment to sustainability, realigning their business models, and integrating environmental practices through cooperation of different stakeholders and innovative approaches. Thus in this chapter, we elucidate the high time for all enterprises to contribute to protecting the environment through developing and implementing innovative sustainable development practices and strategies such as green human resources, circular economy, and human resources analytics. All of which will result in efficient utilization of resources without harming the environment and interests of present and future generations.

1.2 Circular Economy

The concept of Circular Economy (CE) has received great attention among academic scholars throughout the past decade. Companies have also witnessed an increased awareness of the potential opportunities offered by CE in relation to value creation for themselves and their different stakeholders (EMF, 2015). The below Table 1 presents different conceptualizations of CE.

Although Masi et al. (2018) claimed that there is no comprehensive definition of CE, they have classified three clusters of CE, specifically "CE as a new label for an existing concept," "CE as a prescriptive set of existing concepts and practices," and "new definitions that integrate economic, environmental, and social considerations." Within this perspective, the European Environment Agency in 2016 developed five

Table 1 Definitions of circular economy

Authors/date of publication	Definition of CE
Bocken et al. (2016)	"Design and business model strategies [that are] slowing, closing, and narrowing resource loops" (P: 309).
Yuan et al. (2006)	"The core of [CE is the circular flow of materials and the use of raw materials and energy through multiple phases" (P: 5).
Geng and Doberstein (2008, p. 231)	"Realization of [a] closed loop material flow in the whole economic system" (P: 231).
Ellen MacArthur Foundation (2015)	"An industrial economy that is restorative or regenerative by intention and design" (P: 14).
Webster (2015)	"a circular economy is one that is restorative by design, and which aims to keep products, components and materials at their highest utility and value, at all times" (P: 16).
Lieder & Rashid (2016)	Explained CE from different perspectives: Scarcity of resources, such as energy and material consumption; environmental effects, such as pollution, solid waste; and economic gains, such as increased revenues and reduced cost.
Geissdoerfer et al. (2017)	Incorporated CE into sustainability, with the latter considered as the basic concept and CE as the practical application.
Korhonen et al. (2018)	Stressed that understanding what CE is, it is crucial to comprehend the holistic attribute applied to man–nature relations as the main element. Other aspects of CE are derived from different areas of knowledge.
Reike, et al. (2018)	Indicated that the roots of CE emerged from the concerns of residues of the industrial production processes.
D'amato, et al. (2017)	Advocated that CE emerges from the principles of sustainability with a special focus on urban development and inter-governmental relations in supply chain.
Elia, et al. (2017)	Emphasized specific metrics for evaluating processes that use the concept of CE.
Govindan & Hasanagic (2018)	Reported the use of CE in urban development, emphasizing the integration of the public administration legislations in the circularity processes.

features for CE; firstly, less input and use of natural resources, secondly, increased share of renewable and recyclable resources and energy, thirdly, reduced emissions, fourthly, fewer material losses/residuals, and finally, keeping the value of products, components, and materials in the economy.

The existing literature have acknowledged the adoption of CE results in fewer environmental emissions, decrease in resource consumption, a privileged business competitive position (Masi et al., 2018), and the minimization of risks (Rizos et al., 2016). On one hand, a study by Jonker et al. (2017) on Dutch corporations reported that those firms, which utilized CE business models have significantly reduced usage of raw materials and cost of energy consumption and increased recycling, reusing, and repairing of products. On the other hand, a study by three centers for Business and Environment concluded that as a result of technological developments, CE would create huge savings in resources and externalities, such as health impacts from air pollution (Ellen Mac Arthur Foundation, 2015).

Finally, empirical research have recognized a set of factors as enablers or impediments to the successful implementation of CE. For instance, incorporating an organizational culture conducive to environmental protection, conducting cost/ benefit assessment of CE taking into account the business and the processes risks and opportunities, and resistance to change due to lack of CE knowledge and responsibilities (Liu and Bai, 2014).

2 Achieving Sustainable Development through Green HRM

2.1 Green HRM System Components

The concept of GHRM emerged in the 1990s and received global acceptance in the 2000s (Lee, 2009). Green HRM addresses Green HRM policies, green philosophies, and green practices for managing environmentally related issues and enhancing employees' awareness about green environmental responsibilities, accountability, and engaging in green practices/initiatives.

From a sustainable development perspective, the concept of Green HRM seeks to explain the need for a balance between wealth creation and preserving the environment to enhance sustainability through Green HRM initiatives (Daily & Huang, 2001). Scholars (Halawi & Zaraket, 2018; Bhutto & Auranzeb, 2016) emphasized that there is a growing need for achieving a sustainable environment through Green HRM practices/systems such as recruitment and selection, learning and development, performance management, compensation, and rewards. Bombiak and Kluska (2018) found a positive relationship between friendly environment HR activities and sustainable development, supporting Deshwal (2015) in his study that green HRM reduces environmental waste, restores HR tools and procedures, resulting in lower costs.

The starting point to implement environmentally sustainable goals commences with green recruitment and selection in terms of attracting and hiring candidates with environmental awareness and integrating green awareness in different selection processes, such as paperless interviews and job descriptions (Yusoff & Nejati, 2017, Renwick et al., 2013; Jabbour & Santos, 2008). Green employer branding is an effective method in attracting potential candidates with environmental beliefs as they easily identify with corporations with positive environmental image (Jabbour et al., 2013a, 2013b; Saeed et al., 2018).

Baumgartner and Winter (2014) claimed that green learning and development increase employees' perception of engaging in environmentally friendly behavior. Programs that target improving employees' capabilities on environment management initiatives strengthen employees' ability to respond and adapt to environmental changes. Daily et al. (2012) added that green learning and development will help employees develop a pro-active attitude toward reducing environmental wastage and problems.

Saeed et al. (2018) mentioned that green performance management can be reflected in incorporating green standards into employees' job objectives to engage in environmental activities and hold them accountable for environmental management performance. Compensation and rewards are a significant Green HRM practices for rewarding employees for their eco/environmental initiatives and performance. Scholars such as Veleva and Ellenbecker (2001) and Haque (2017) established a significant positive relationship between green compensation and commitment to environmental performance. Rewards whether monetary or nonmonetary incentives are equally powerful in promoting environmentally friendly behavior.

In concluding remarks, corporations integrate Green HRM practices into their business, such as recruiting people with environmental conscientiousness, motivating employees to learn and engage in pro-environmental behaviors (Tseng et al., 2013; Cherian & Jacob, 2012; Djellal & Gallouj, 2016), rewarding innovative environmental performance, and encouraging green involvement (Saeed et al., 2018) will ease pressure on the environment and scarce resources and reduce wastage and operational costs.

2.2   Green HRM and Circular Economy

The link between Green HRM and CE is captured in the integrative framework proposed by Jabbour et al. (2019). The framework aligns Green HRM systems and green HR organizational enablers with the CE business model—RESOLVE and their ultimate impact on firms' sustainable performance.

The Green HRM systems include environmental recruitment and selection, eco-focused training, eco-aware performance assessment and rewards, and the green HR organizational enablers such as ecological organizational culture, green teams, and eco-focused employee empowerment, all of which contribute to the firm's green strategy to enable a CE.

The CE business model—RESOLVE stands for REgenerate, share, optimize, loop, virtualize, and exchange. The *REgenerate* component refers to renewable energy and materials. Regenerate health of ecosystems and return recovered biological resources to biosphere. The *Share* element entails maximum utilization of products via sharing them among users, reusing them throughout their technical lifetime and prolonging their life through repair and maintenance. The *Optimize* factor encompasses performance improvement and efficiency of a product, removal of production, and supply chain waste. The *Loop* as the name implies keeps materials in closed loops and prioritizes inner loops, remanufacturing infinite products and recycling materials. The *Virtualize* component is concerned with delivering utility virtually. The *Exchange* as it is implied deals with replacing old materials with advanced nonrenewable materials, choose new products, and apply new technology (Jabbour et al., 2019).

The integrative framework conceptualizes the potential interface between GHRM and CE business models, and insights of the CE phenomenon among academics. Jabbour et al. (2019) concluded that this framework adds a new and fresh perspective to debates around the development of CE business models through its incorporation of a green human side. They laid the foundation for scholars to build a firm theoretical and practical integrated view of Green HRM and CE.

2.3   Green HRM and Environmental Performance

The concept of green management for sustainable development in the extant literature has revealed numerous definitions; all of which explains the need for a balance between organizations' growth and preserving the environment for achieving long terms well-being, prosperity of future generations (Daily & Huang, 2001). The literature has presented ample studies showing how organizations can achieve environmental sustainability through Green HRM practices.

Green HRM practices such as green recruitment and selection, eco-focused training, green performance management and rewards, promoting eco-friendly culture, green teams, and employee empowerment positively impact environmental initiatives such as low carbons eco-innovations (Jabbour et al., 2015 & Renwick et al., 2016). Moreover, other studies (Gholami et al., 2016; Guerci & Pedrini, 2014; Renwick et al., 2013; Zibarras & Coan, 2015) asserted a positive relationship between Green HRM and green sustainability goals.

Renwick et al. (2013) also found a link between Green HRM workplace practices and organizational environmental goals. Along the same thoughts, Daily et al. (2012) concluded that green training and employee empowerment specifically are considered as a key enablers of environmental sustainability. This is supported by Subramanian et al. (2019) affirming that Green HRM practices can enhance firms' sustainability performance in terms of achieving better environmental performance.

Kim et al. (2019) reported that the positive impact of Green HRM on a firm's sustainability results in the fact that Green HRM can enhance employees' green behavior. Other researchers (Ramus & Ulrich, 2000; Graves et al., 2013) also found a connection between employee motivation for sustainability, eco-innovation, and pro-environmental behaviors. As indicated in the HRM literature (Jabbour and Santos, 2008; Fernandez et al., 2017) Green HRM can advance sustainability-related initiatives such as eco-design, environmental management systems, and low-carbon management. From the outset, we can conclude that Green HRM practices have a substantial impact on promoting environmental performance at organizational level, consequently leading to sustainable performance.

3   Role of HR Analytics

3.1   HR Analytics: Concept Definition

Big data has become essential to transform businesses and to improve their potential to respond to environmentally sustainable business performance (El-Kassar & Singh, 2019). Although the use of analytics has a long history associated with human capital management (HCM) decisions, many organizations still rely on these tools for reporting simple descriptive statistics and correlations. Yet, advanced analytics are not yet fully utilized in HR decisions (Sesil, 2014). The concept of HR analytics is driven by technological advancements and the availability of Human resource information systems (HRIS) that can provide big HR data (Van den Heuvel & Bondarouk, 2017). Lawler et al. (2004) study provided an initial attempt to explore the concept. Authors distinguish between HR analytics and HR metrics. They argue that while HR metrics are measures of key HR-related outcomes, HR analytics embrace a more complex statistical approach used by companies in order to show the impact of HR activities on organizational outcomes.

In their review of HR analytics literature, Marler and Boudreau (2017), a consensus was found on the definition of HR analytics. For example, while Aral et al. (2012) view was only limited to applying HR analytics in organizations as a way to measure and monitor individual performance, Ulrich and Dulebohn (2015) informed us that HR analytics includes rigorous track of HR investments and outcomes. Moreover, Van den Heuvel and Bondarouk (2017) defined the concept as "a systematic identification and quantification of the people-drivers of business outcomes, with the purpose of making better decisions" (Van den Heuvel & Bondarouk, 2017, p. 160). Given these commonalities, Marler and Boudreau (2017) have defined HR analytics as an *"An HR practice enabled by information technology that uses descriptive, visual, and statistical analyses of data related to HR processes, human capital, organizational performance, and external economic benchmark to establish business impact and enable data driven decision-making"* (Marler & Boudreau, 2017, p. 15). Since the definition is broad and was built upon a comprehensive review, it will form the basis of the current research.

Overall, HR analytics enable strategic HR decisions that are based on HR metrics and analytics that are largely supported by the best available scientific evidences.

This requires building HR analytics capabilities and the use of HR data that are collected through primary and secondary research or through mining the organizations' data available in its Human Resource Information system (HRIS) (Falletta & Combs, 2020). Then, data can be transformed into meaningful insights through specialized software (like Oracle, SAP, and SAS) that incorporates HR analytical capabilities such as predictive analytics, process analytics, and real-time analytics (Strohmeier et al., 2015). Finally, HR analytics results are to be effectively communicated (through scorecards and dashboards) and used to enable HR strategy creation and to inform HR evidence-based decision-making.

3.2   HR Analytics Elements

Building upon the above definition, it becomes obvious the role HR analytics play to connect HR programs and practices to organizational outcomes. Boudreau and Ramstad (2006) introduced the LAMP model that describes the critical components of HR analytics (the letters stand for logic, analytics, measures, and process) that can be used as a basis to uncover the relationship between HRM processes and practices and business outcomes.

The "*logic*" element provides a framework that emphasizes the connection between talent management and the strategic success of organizations. For example, it is logical to assume a link between employees' health and turnover and absenteeism rates. Many companies today are relying on HR analytics to evaluate which characteristics are related to good employees, and this information is used to help with employee selection (Sesil, 2014). In other words, decision makers should understand how HR information could enhance their HR decisions (Cascio & Boudreau, 2015).

The "*Analytics*" component encompasses the tools and techniques used to transform HR data into relevant insights. This component emphasizes engaging analysis that tests relationships between measures and outcomes (Kryscynski et al., 2018). Analytics also involves sourcing the right data both from within and outside the HR functions. For effective analysis, the right skills and competences are required to enable HR professionals to identify the key issues in the data (Boakye & Lamptey, 2020).

The "*Measures*" component includes the numbers and indices calculated for data systems. Boudreau and Ramstad (2006) developed the HC Bridge framework to uncover HR measures that includes efficiency measures (use of resources to implement HR activities), effectiveness measures (the effect of HR activities on employees), and impact measures (the effect of the actions on organizational performance). Marler and Boudreau (2017) concluded in their review that the LAMP model operationalizes aspects of the HR scorecard, introduced by Kaplan and Norton (1992), as part of HR analytics in organizations. The HR scorecard model links HRM processes and people to business outcomes. Lawler and Boudreau (2015) reported that the HR scorecard is one of the most frequent analytics elements used by HR leaders and is listed as existing now in organizations.

Finally, the *"process"* component refers to the communication and knowledge transfer mechanisms used to make HR information accessible and used by decision makers in organizations (Marler & Boudreau, 2017). It explains how measurement can affect the decisions and behaviors of an organization that occurs within complex webs of social structures, organizational cultural norms, and knowledge frameworks (Cascio & Boudreau, 2010). Boudreau and Ramstad (2006) suggested that these four components can be used to explain the link between HRM processes and strategies and organizational performance.

3.3   Strategic Role of HR Analytics: Applying the Resource-Based View

The strategic human resource management (SHRM) field stresses the role of HR activities and systems in supporting business strategy (Wright & McMahan, 1992). In order to provide a proper explanation for the strategic value of HRM, the resource-based view (RBV) was adopted. The RBV is one perspective that provided a rationale for how a firm's human resources could provide a potential source of sustainable competitive advantage (Boselie et al., 2005). According to the RBV, human resources, as part of the company's internal resources, could be viewed as sources of competitive advantage. Additionally, HR delivers "added value" through the strategic development of the company's important resources that are rare, inimitable, and non-substitutable. This assumption has brought legitimacy to HR's assertion that people are strategically important to improve organizational performance (Wright et al., 2001). The RBV is heavily used in the Strategic HRM literature both theoretically and empirically as a rationale for the HRM performance link (e.g., Arthur, 1994; Huselid, 1995).

HR analytics, through their four components, can be a key to understanding the cause–effect relationship between HRM processes and strategic HRM and business outcomes. Particularly, HR analytics can cause better performance and may lead to a competitive advantage when it is unique and value producing (Marler & Boudreau, 2017). Aral et al. (2012) study found that HR analytics can predict company-level productivity when it is combined with HCM software and pay for performance compensation. The results are consistent with the LAMP model in that it was a combination of *Analytics* (HR analytics) with *process* (Pay for Performance) and *measures* (HCM software) that produces this strategic value (productivity). Another study by Harris et al. (2011) provides examples to link six analytical tools that comprise HR analytics with business performance. For example, they gave an example of Google that uses HR analytics to predict employee performance using their applicant database. Another example is Sysco that utilized HR analytics to establish causal relationship between work climate surveys, delivery driver employee satisfaction, customer loyalty, and higher revenue.

The strategic value of HR analytics can also be viewed from the well-known AMO (Ability-Motivation-Opportunity) model proposed by Bailey (1993). The model suggested that ensuring the employee's discretionary effort needed three components: employees should have the skills needed to perform the task, employees have to be motivated to perform the task, and employers had to offer them the opportunity to participate (Appelbaum et al., 2000). The framework is heavily used in the HRM literature that examined the link between HRM and different measures of employee and organizational performance. In particular, the model is initially introduced in the HRM literature by Appelbaum et al. (Appelbaum et al., 2000) when they conceptualized HR practices in terms of its ability to enhance together employee performance: individual ability (A), motivation (M), and the opportunity to participate (O) (Boselie et al., 2005; Obeidat et al., 2020). The use of the AMO model as a way to conceptualize HRM components is theoretically plausible (Jiang et al., 2013) and empirically valid (Obeidat et al., 2020).

Applying the AMO model to Green HRM, Green HRM includes practices that improve employees' Green knowledge, skills, and abilities (KSA) inventory, which are called ability-enhancing practices (Wright & Kehoe, 2008). For example, HR practices such as the formal staffing and systematic training practices influence employees' Green ability to perform their tasks. Moreover, HR practices such as providing adequate incentives to employees who are green aware and the adoption of a green objective performance appraisal (also called motivation-enhancing practices) promote employees' green motivation. Finally, opportunity-enhancing HR practices such as quality circles, flexible work arrangement, and effective communication tools, promotes sustainable practices and provides opportunities for employees to participate in green initiatives (Obeidat et al., 2020; Boxall & Purcell, 2003; Marin-Garcia & Tomas, 2016). This could help in understanding what proper HR metrics are needed to measure Green HRM functions and practices that promote ability, motivation, and opportunity.

3.4   Green HRM and HR Metrics

HR metrics were designed to measure HR efficiency. Many HR metrics as shown in Table 2 were developed to determine the value and effectiveness of HRM processes and practices in areas such as turnover, training, return on human capital, costs of labor, and expenses per employee. Organizations use HR metrics as benchmarks of efficiency, effectiveness, and impact that resulted in competitive advantages (Handa, 2014). Linking HR metrics with HR analytics, it was argued that HR analytics transforms HR data and HR Metrics (Measures) by using statistics and research to provide accurate insights on business performance (Fitz-enz, 2010).

Table 2 Examples of HR metrics

Metrics	How to calculate it...
Time to hire	The number of days between a candidate applying for a job, and then accepting a job offer
Cost per hire	Total cost of hiring/the number of new hires
Early turnover	Percentage of recruits leaving in the first year
Turnover rate	The percentage of employees who left a company over a certain period of time
Training cost per employee	Training budget divided by the number of employees available for training
Skill development achieved per training program	How effective was the training program in helping employees gain relevant knowledge and skills
Payroll cost	All costs incurred by an employer to compensate its employees

HR Metrics can help improve decisions in Green HRM. For example, an important metric in recruitment is "Cost per hire." It measures how much it costs the company to hire a new employee. With green recruitment, organizations can attract individuals with a more positive environmental attitude, who are willing to work for sustainable organizations (Bohlmann et al., 2018). Accordingly, organizations with high level of environmental performance are more capable of attracting valuable talents. Candidates with green attitudes are more willing to accept a job offer from green organizations. This means that green recruitment can lead to decreasing "Cost Per Hire" (Table 2)

3.5 Linking HR Analytics with Green HRM and Circular Economy

Building upon the view that people are the most significant factor affecting business performance, HR analytics enables quantifying, evaluating, and controlling human behavior critical for improved workforce performance (Momin & Mishra, 2016; Nienaber & Sewdass, 2016). Likewise, HR analytics can provide timely and relevant data that could help in applying effective Green HRM practices. Big data has enough potential to influence environmental practices in organizations (Keeso; 2014). Furthermore, it is argued that big data can assist in achieving organizational goals that focus on improving social and environmental sustainability (Song et al., 2017). El-Kassar and Singh (2019) argue that integrating Green HRM with big data management can develop sustainable capabilities that could lead to better sustainable performance.

The possible link between big data analytics and Green HRM and circular economy can be viewed from the dynamic capability view (DCV). DCV refers to the firm's ability to address and adapt to changing environments through proper integration and reconfiguration of internal and external resources (Teece et al., 1997). Organizations can improve their abilities in a dynamic environment through dynamic capabilities, which ultimately can lead to improved sustainable performance (Edwin Cheng et al., 2021). Faced with internal and external pressure to adapt and implement environmentally friendly business activities, it is becoming increasingly vital for organizations to develop and apply green organizational capabilities. Previous work has conceptualized big data analytics as an organizational resource associated with building dynamic capabilities (El-Kassar & Singh, 2019).

In specific, big data analytics provides insightful information on concepts like circular economy and Green HRM (Jia et al., 2018) since it creates useful information for decision-making in several circular economy-based activities to integrate processes and share resources (Jabbour et al., 2019). These insights can then serve as a building block for making important decisions regarding green management practices. Levenson (2018) suggests that the use of analytics and metrics provide great potential for the improvement of the quality of decision-making in HRM issues in organizations. Levenson's study argued that the metrics are important in increasing the levels of insightful, analytically based decision-making in HRM.

Many Green HRM practices could be implemented as part of green management practices that improve sustainable performance. For example, in green recruitment and selection, organizations select candidates who are well informed about greening to fill out vacant positions. Also, recruitment posts can include environmental commitment criteria and the messages used could reflect environmental efforts of the company. Job candidates could be asked environmental-related questions during the interview process.

In green training, organizations can consider environmental issues in training needs assessment. Employees can receive environmental awareness training during their tenure. Overall, companies focus on environmental training and give it a priority compared with other training types. In green performance management, green-related criteria are incorporated into the performance appraisal system and employees receive feedback from their managers with regard to environmental goals.

Green compensation is also an important practice to credit employees who adopt green behaviors at work. Specifically, employees can receive financial rewards for environmental achievements and for creative environmental initiatives. Moreover, incentives are given to encourage environmentally friendly activities. Green empowerment practices are also important to achieve sustainable outcomes. Employees have the opportunities to participate in green suggestion schemes and to consult with management about solving environmentally related problems.

Overall, HR analytics can serve as a tool for measuring the level of effectiveness and efficiency of the implemented Green HRM practices. Specifically, organizations may have policies to increase employees who are green aware. In this context, HR analytics can help identify the set of potential measures that may increase candidates with environmental awareness during the recruitment and selection process. Also, their future impact on organizational performance can be assessed (such as its effect on return on investment).

In *green training and development*, the HR team can use analytics-based dashboards to develop personalized green training plans for new employees. Usually, dashboards are used to calculate the return on the training investment. Therefore, the HR department can adopt similar techniques to calculate the effect of green training on organizational performance, like financial and environmental performances. HR analytics can help in implementing proper *green performance management and green compensation* practices. In particular, HR analytics can provide the tools to link individual performance appraisal results (that contain environmental criteria) to compensation with the result that can be linked to organizational performance.

Green empowerment allows employees to be engaged in participating in green suggestion schemes and will have the opportunity to negotiate with management about green workplace agreements. The HR team can collect data about green engagement rate and design the support employees may need to enhance their engagement. Moreover, these data can help senior management to understand the drivers of employee turnover and to design policies to improve their retention.

4 Conclusion

In order to achieve sustainable development goals in organizations, the term Green HRM is here to stay. Particularly, Green HRM practices and functions can help organizations today improve their economic, social, and environmental performances.

However, for organizations trying to promote the Circular economy model through implementing green practices, such as Green HRM, implementing only good Green practices are not enough. In order to strengthen the link between Green HRM and sustainable performance, the adoption of big data analytics becomes necessary. HR analytics help to create a good decision on the type of Green HRM practices that needs to be implemented to achieve environmental goals.

The other central message of this chapter is the need for a more integrative framework to investigate the possible link between Green HRM and sustainable performance by taking into consideration other important factors that may affect this link, such as HR analytics. In this regard, considering the original theoretical underpinnings of HRM has considerable merit. Particularly, employing HR theoretical foundations such as the RBV and the AMO model can help to better understand this link. Reflecting on this, we proposed a conceptual framework that could guide future research on Green HRM, HR analytics, and sustainable performance link.

A conceptual Framework

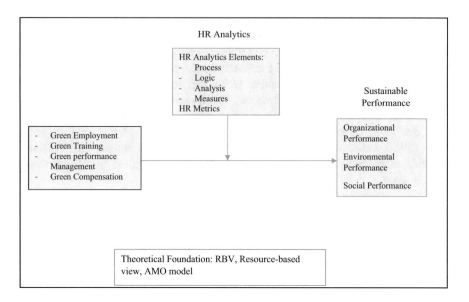

References

Appelbaum, E., Bailey, T., Berg, P., & Kalleberg, A. (2000). *Manufacturing advantage: Why high-performance work systems pay off.* Cornell University Press.

Aral, S., Brynjolfsson, E., & Wu, L. (2012). Three-way complementarities: Performance pay, human resource analytics, and information technology. *Management Science, 58*(5), 913–931.

Arthur, J. B. (1994). Effects of human resource systems on manufacturing performance and turnover. *Academy of Management Journal, 37*(3), 670–687.

Bailey, T. (1993). Organizational innovation in the apparel industry. *Industrial Relations: A Journal of Economy and Society, 32*(1), 30–48.

Baumgartner, R., & Winter, T. (2014). The sustainability manager: A tool for education and training on sustainability management. *Corporate Social Responsibility and Environment Management, 21*(3), 167–174.

Bhutto, S., & Auranzeb, Z. (2016). Effects of green human resource management on firm performance: An empirical study on Pakistani firms. *European Journal of Business and Management, 8*(16), 119–125.

Boakye, A., & Lamptey, Y. A. (2020). The rise of HR analytics: Exploring its implications from a developing country perspective. *Journal of Human Resource Management, 8*(3), 181–189.

Bocken, N., de Pauw, I., Bakker, C., & van der Grinten, B. (2016). Product design and business model strategies for a circular economy. *Journal of Industrial and Production Engineering, 33*, 308–320.

Bohlmann, C., Krumbholz, L., & Zacher, H. (2018). The triple bottom line and organizational attractiveness ratings: The role of pro-environmental attitude. *Corporate Social Responsibility and Environmental Management, 25*(5), 912–919.

Bombiak, E., & Kluska, M. (2018). Green human resource management as a tool or the sustainable development of enterprises: Polish young company experience. *Sustainability, MDPI, 10*(6), 1–22.

Boselie, P., Dietz, G., & Boon, C. (2005). Commonalities and contradictions in HRM and performance research. *Human Resource Management Journal, 15*(3), 67–94.

Boudreau, J., & Cascio, W. (2017). Human capital analytics: Why are we not there? *Journal of Organizational Effectiveness: People and Performance, 4*(2), 119–126. https://doi.org/10.1108/JOEPP-03-2017-0021

Boudreau, J., & Ramstad, P. M. (2006). Talentship and HR measurement and analysis: From ROI to strategic organizational change. *Human Resource Planning., 29*, 25–33.

Boxall, P., & Purcell, J. (2003). *Strategy and human resource management*. Palgrave Macmillan.

Brown, L. (1991). Bridging organizations and sustainable development. *Human Relations, 44*(8), 807–831.

Cascio, W., & Boudreau, J. (2015). The search for global competence: From international HR to talent management. *Journal of World Business., 51*. https://doi.org/10.1016/j.jwb.2015.10.002

Cascio, W. F., & Boudreau, J. W. (2010). *Investing in people: Financial impact of human resource initiatives* (2nd ed.). Pearson Education.

Cherian, J., & Jacob, J. (2012). Green marketing: A study of consumers' attitude towards environment friendly products. *Asian Social Science, 8*(12), 117–132.

D'amato, D., Droste, N., Allen, B., Kettunen, M., Lahtinen, K., Korhonen, J., Leskinen, P., & Matthies, B. (2017). Green, circular, bio economy: A comparative analysis of sustainability avenues. *Journal of Cleaner Production, 168*, 716–734. https://doi.org/10.1016/j.jclepro.2017.09.053

Daily, B., Bishop, J., & Massoud, J. (2012). The role of training and empowerment in environmental performance: A study of the Mexican maquiladora industry. *International Journal of Operations Production Management, 32*(5), 631–647.

Daily, B. F., & Huang, S. (2001). Achieving sustainability through attention to human resource factors in environmental management. *International Journal of Operations & Production Management, 21*, 1539–1552.

Deshwal, P. (2015). Green HRM: An organizational strategy of greening people. *International Journal of Applied Research, 1*(13), 176–181.

Djellal, F., & Gallouj, F. (2016). Service innovation for sustainability: Paths for greening through service innovation. In *Service innovation* (pp. 187–215). Springer.

Edwin Cheng, T. C., Kamble, S. S., Belhadi, A., Ndubisi, N. O., Lai, K. H., & Kharat, M. G. (2021). Linkages between big data analytics, circular economy, sustainable supply chain flexibility, and sustainable performance in manufacturing firms. *International Journal of Production Research*, 1–15. https://doi.org/10.1080/00207543.2021.1906971

El-Kassar, A.-N., & Singh, S. K. (2019). Green innovation and organizational performance: The influence of big data and the moderating role of management commitment and HR practices. *Technological Forecasting and Social Change, Elsevier, 144*(C), 483–498.

Elia, V., Grazia, M., & Tornese, G. (2017). Measuring circular economy strategies through index methods: A critical analysis. *Journal of Cleaner Production, 142*(4), 2741–2751. https://doi.org/10.1016/j.jclepro.2016.10.196

Ellen MacArthur Foundation. (2015). *Growth within: A circular economy vision for a competitive Europe*. Ellen MacArthur Foundation.

Falletta, S., & Combs, W. (2020). The HR analytics cycle: A seven-step process for building evidence-based and ethical HR analytics capabilities. *Journal of Work-Applied Management*. https://doi.org/10.1108/JWAM-03-2020-0020

Fernandez, L., et al. (2017). The effect of clean development mechanism projects on human resource management practices in Brazil. *Int. Journal of Operations Production Management, 37*(10), 1348–1365. In Jabbour, C., Sarkis, J., Jabbour, A., Renwick, D., Singh, S., Grebinevych, O., Kruglianskas, I., Filho, M. (2019). Who is in charge? A review and a research agenda on the 'human side' of the circular economy. *Journal of Cleaner Production,* *222,* 793–801. http://www.elsevier.com/locate/jclepro.

Fischer-Kowalski, M., & Swilling, M. (2011). *Decoupling: Natural resource use and environmental impacts from economic growth.* United Nations Environment Program.

Fitz-enz, J. (2010). *The new HR analytics: Predicting the economic value of your company's human capital investments.* AMACOM–Division of American Management Association.

Geissdoerfer, M., Savaget, P., Bocken, N., & Hultink, E. (2017). The circular economy - a new sustainability paradigm? *Journal of Cleaner Production, 143,* 757–768.

Geng, Y., & Doberstein, B. (2008). Developing the circular economy in China: Challenges and opportunities for achieving "leapfrog development". *International Journal of Sustainable Development & World Ecology, 15,* 231–239.

Gholami, H., Rezaei, G., Saman, M., Sharif, S., & Zakuan, N. (2016). State-of-the-art green HRM system: Sustainability in the sports center in Malaysia using a multi-methods approach and opportunities for future research. *Journal of Cleaner Production, 124,* 142–163.

Govindan, K., & Hasanagic, M. (2018). A systematic review on drivers, barriers, and practices towards circular economy: A supply chain perspective. *International Journal of Production Resources, 56,* 278–311.

Graves, L., Sarkis, J., & Zhu, Q. (2013). How transformational leadership and employee motivation combine to predict employee pro-environmental behaviors in China. *Journal of Environmental Psychology, 35,* 81–91.

Guerci, M., & Pedrini, M. (2014). The consensus between Italian HR and sustainability managers on HR management for sustainability-driven change - towards a 'strong' HR management system. *International Journal of Human Resource Management, 25*(13), 1787–1814.

Halawi, A., & Zaraket, W. (2018). Impact of green human resource management on employee behaviour. *Journal of Applied Business Research, 6*(1), 18–34.

Handa, D., 2014. Garima: Human resource (HR) analytics: Emerging trend in HRM (HRM).

Haque, F. (2017). The effects of board characteristics and sustainable compensation policy on carbon performance of UK firms. *The British Accounting Review, 49*(3), 347–364. https://doi.org/10.1016/j.bar.2017.01.001

Harris, J. G., Craig, E., & Light, D. A. (2011). Talent and analytics: New approaches, higher ROI. *Journal of Business Strategy, 32*(6), 4–13. https://doi.org/10.1108/02756661111180087

Huselid, M. A. (1995). The impact of human resource management practices on turnover, productivity, and corporate financial performance. *The Academy of Management Journal, 38*(3), 635–672.

Jabbour, C., & Santos, F. (2008). Relationships between human resource dimensions and environmental management in companies: Proposal of a model. *Journal of Cleaner Production, 16*(1), 51–58.

Jabbour, C., de Sousa Jabbour, A., Govindan, K., Teixeira, A., & de Souza Freitas, W. (2013b). Environmental management and operational performance in automotive companies in Brazil: The role of human resource management and lean manufacturing. *Journal of Cleaner Production, 47,* 129–140. https://doi.org/10.1016/jjclepro.2012.07.010

Jabbour, C., Neto, A., Gobbo, J., de Souza Ribeiro, M., & de Sousa Jabbour, A. (2015). Eco-innovations in more sustainable supply chains for a low-carbon economy: A multiple case study of human critical success factors in Brazilian leading companies. *International Journal of Production Economy, 164,* 245–257.

Jabbour, C., Sarkis, J., Jabbour, A., Renwick, D., Singh, S., Grebinevych, O., Kruglianskas, I., & Filho, M. (2019). Who is in charge? A review and a research agenda on the 'human side' of the circular economy. *Journal of Cleaner Production, 222,* 793–801. http://www.elsevier.com/locate/jclepro

Jabbour, C. J. C., de Sousa Jabbour, A. B. L., Govindan, K., Teixeira, A. A., & de Souza Freitas, W. R. (2013a). Environmental management and operational performance in automotive companies in Brazil: The role of human resource management and lean manufacturing. *Journal of Cleaner Production, 47,* 129–140. https://doi.org/10.1016/j.jclepro.2012.07.010

Jia, J., Liu, H., Chin, T., & Hu, D. (2018). The continuous mediating effects of GHRM on employees' green passion via transformational leadership and green creativity. *Sustainability, 10*(9), 3237.

Jiang, K., Takeuchi, R., & Lepak, D. P. (2013). Where do we go from here? New perspectives on the black box in strategic human resource management research. *Journal of Management Studies, 50*(8), 1448–1480.

Jonker, J., Stegeman, H., Faber, N. (2017). The circular economy - developments, concepts, and research in search for corresponding business models. White paper. https://www.researchgate.net/publication/313635177

Kaplan, R. S., & Norton, D. P. (1992). The balanced scorecard - measures that drive performance. *Harvard Business Review, 70*, 172.

Kim, Y., Kim, W., Choi, H., & Phetvaroon, K. (2019). The effect of green human resource management on hotel employees' eco-friendly behavior and environmental performance. *International Journal of Hospitality Management, 76*, 83–93.

Korhonen, J., Honkasalo, A., & Seppala, J. (2018). Circular economy: The concept and its limitations. *Ecological Economics, 143*, 37–46.

Kryscynski, D., Reeves, C., Stice-Lusvardi, R., Ulrich, M., & Russell, G. (2018). Analytical abilities and the performance of HR professionals. *Human Resource Management, 57*, 715–738. https://doi.org/10.1002/hrm.21854

Lawler, E. E., Levenson, A. and Boudreau, J. (2004), "HR metrics and analytics uses and impacts", working paper, CEO Publication. Retrieved December 7, 2015, from http://classic.marshall.usc.edu/assets/048/9984.pdf

Lawler, E., & Boudreau, J. (2015). *Global trends in human resource Management: A twenty-year analysis*. Stanford University Press. https://doi.org/10.1515/9780804794558

Lee, K. (2009). Why and how to adopt green management into business organizations? The case study of Korean SMEs in manufacturing industry. *Management Decision, 47*(7), 1101–1121.

Levenson, A. (2018). Using workforce analytics to improve strategy execution. *Human Resource Management, 57*(3), 685–700.

Lieder, M., & Rashid, A. (2016). Towards circular economy implementation: A comprehensive review in context of manufacturing industry. *Journal of Cleaner Production, 115*, 36–51.

Liu, Y., & Bai, Y. (2014). An exploration of firms' awareness and behavior of developing circular economy: An empirical research in China. *Resources, Conservation and Recycling, 87*, 145–152.

Margherita, A. (2021). Human resources analytics: A systematization of research topics and directions for future research. *Human Resource Management Review.*, 100795. https://doi.org/10.1016/j.hrmr.2020.100795

Marin-Garcia, J. A., & Tomas, J. M. (2016). Deconstructing AMO framework: A systematic review. *Intangible Capital, 12*(4), 1040–1087.

Marler, J. H., & Boudreau, J. W. (2017). An evidence-based review of HR analytics, the. *International Journal of Human Resource Management, 28*(1), 3–26. https://doi.org/10.1080/09585192.2016.1244699

Masi, D., Kumar, V., Garza-Reyes, J., & Godsell, J. (2018). Towards a more circular economy: Exploring the awareness, practices, and barriers from a focal firm perspective. *Production Planning & Control, 29*(6), 539–550.

Momin, W. Y. M., & Mishra, K. (2016). HR analytics: Re-inventing human resource management. *International Journal of Applied Research, 2*(5), 785–790.

Nienaber, H., & Sewdass, N. (2016). A reflection and integration of workforce conceptualisations and measurements for competitive advantage. *Journal of Intelligence Studies in Business, 6*(1), 5–20.

Obeidat, S. M., Al Bakri, A. A., & Elbanna, S. (2020). Leveraging "Green" human resource practices to enable environmental and organizational performance: Evidence from the Qatari Oil and Gas Industry. *Journal of Business Ethics, 164*, 371–388. https://doi.org/10.1007/s10551-018-4f075-z

Ramus, C., & Ulrich, S. (2000). The roles of supervisory support behaviors and environmental policy in employee "Ecoinitiatives" at leading-edge European companies. *Academy of Management Journal, 43*(4), 605–626.

Reike, D., Vermeulen, W., & Witjes, S. (2018). The circular economy: New or refurbished as CE 3.0? Exploring controversies in the conceptualization of the circular economy through a focus on history and resource value retention options. *Resources, Conservation & Recycling, 135*, 246–264. https://doi.org/10.1016/j.resconrec.2017.08.027

Renwick, D., Redman, T., & Maguire, S. (2013). Green HRM: A review and research agenda. *International Journal of Management Review, 15*(1), 1–14.

Renwick, D., Jabbour, C., Muller-Camen, M., Redman, T., & Wilkinson, A. (2016). Introduction: Contemporary developments in green (environmental) HRM scholarship. *International Journal of Human Resource Management, 27*(2), 1–16.

Rizos, V., Behrens, A., Vander Gaast, W., Hofman, E., Ioannou, A., Kafyeke, T., Flamos, A., Rinaldi, R., Papadelis, S., Garbers, M., & Topi, C. (2016). Implementation of circular economy business models by small and medium-sized enterprises (SMEs): Barriers and enablers. *Sustainability, 8*(11), 1212. https://doi.org/10.3390/su8111212

Saeed, B., Afsar, B., Hafeez, S., Khan, I., Tahir, M., & Afridi, M. (2018). Promoting employee's proenvironmental behavior through green human resource management practices. *Corporate Social Responsibility and Environmental Management, 26*(2), 424–438. https://onlinelibrary.wiley.com/doi/abs/10.1002/csr.1694

Sesil, J. C. (2014). *Applying advanced analytics to HR management decisions.* Pearson Education, Inc.

Song, M., Cen, L., Zheng, Z., Fisher, R., Liang, X., Wang, Y., & Huisingh, D. (2017). How would big data support societal development and environmental sustainability? Insights and practices. *Journal of Cleaner Production, 142*, 489–500.

Strohmeier, S., Piazza, F., & Neu, C. (2015). Trends der human resource intelligence und analytics. In *Human resource intelligence und analytics* (pp. 339–367). Springer. https://doi.org/10.1007/978-3-658-03596-9_11

Subramanian, N., Roscoe, S., & Jabbour, C. (2019). Green human resource management and the enablers of green organizational culture: Enhancing a firm's environmental performance for sustainable development. In *Business strategy and the environment.*

Teece, D. J., Pisano, G., & Shuen, A. (1997). Dynamic capabilities and strategic management. *Strategic Management Journal, 18*(7), 509–533.

Tseng, M., Tan, R., & Siriban-Manalang, A. (2013). Sustainable consumption and production for Asia: Sustainability through green design and practice. *Journal of Cleaner Production, 40*, 1–5. https://doi.org/10.1016/j.jclepro.2012.07.015

Ulrich, D., & Dulebohn, J. H. (2015). Are we there yet? What's next for HR? *Human Resource Management Review, 25*(2), 188–204.

United Nations. (1992). Report of the United Nations Conference on Environment and Development (Vol. I). https://undocs.org/en/A/CONF.151/26/Rev.1

United Nations. (2015). *Transforming our world: The 2030 agenda for sustainable development.* https://www.un.org

van den Heuvel, S., & Bondarouk, T. (2017). The rise (and fall?) of HR analytics: A study into the future application, value, structure, and system support. *Journal of Organizational Effectiveness: People and Performance., 4*. https://doi.org/10.1108/JOEPP-03-2017-0022

Veleva, V., & Ellenbecker, M. (2001). Indicators of sustainable production: Framework and methodology. *Journal of Cleaner Production, 9*(6), 519–549. https://doi.org/10.1016/S0959-6526(01)00010-5

Webster, K. (2015). *The circular economy: A wealth of flows.* Ellen MacArthur Foundation, Isle of Wight.

Wright, P. M., & Kehoe, R. R. (2008). Human resource practices and organizational commitment: A deeper examination. *Asia Pacific Journal of Human Resources, 46*(1), 6–20.

Wright, P. M., & McMahan, G. C. (1992). Theoretical perspectives for strategic human resource management. *Journal of Management, 18*(2), 295–320.

Wright, P., Gardner, T., Moynihan, L., Park, H., Gerhart, B., & Delery, J. (2001). Measurement Error in Research on Human Resources and Fi*rm* Performance: Additional Data and Suggestions for Future Research. In *CAHRS Working Paper Series* (Vol. 54). https://doi.org/10.1111/j.1744-6570.2001.tb00235.x

Yuan, Z., Bi, J., & Moriguichi, Y. (2006). The circular economy: A new development strategy in China. *Journal of Industrial Ecology, 10*, 4–8.

Yusoff, Y., & Nejati, M. (2017). A conceptual model of green HRM adoption towards sustainability in hospitality industry. In *Driving Green Consumerism Through Strategic Sustainability Marketing*. IGI Global.

Zibarras, L., & Coan, P. (2015). HRM practices used to promote pro-environmental behavior: A UK survey. *International Journal of Human Resource Management, 26*(16), 2121–2142.

Understanding Malaysian B40 Schoolchildren's Lifestyle and Educational Patterns Using Data Analytics

Puteri N. E. Nohuddin ⓘ**, Zuraini Zainol** ⓘ**, Marja Azlima Omar** ⓘ**, Hanafi Al Hijazi** ⓘ**, and Nora Azima Noordin** ⓘ

1 Introduction

The Malaysian government places a high value on education for all of its citizens; it is one of the 11th RMK's top goals. One of the initiatives focuses on education and children, notably in the B40 community. To maintain the country's economy thriving, it is necessary to increase people's economic standards. As a result, in order to have a better future, B40 community youth must excel in school or college. Academic monitoring is an important effort that both the school and the parents must make in order to keep track of their children's academic progress. To manage the

P. N. E. Nohuddin (✉)
Faculty of Business, Sharjah Women's Campus, Higher Colleges of Technology, Sharjah, UAE

Institute of IR4.0, National University of Malaysia, Bangi, Malaysia
e-mail: pnohuddin@hct.ac.ae; puteri.ivi@ukm.edu.my

Z. Zainol
Department of Computer Science, National Defence University Malaysia, Kuala Lumpur, Malaysia
e-mail: zuraini@upnm.edu.my

M. A. Omar
Faculty of Social Sciences and Humanities, University of Malaysia of Sabah, Kota Kinabalu, Sabah, Malaysia
e-mail: mazlima@ums.edu.my

H. Al Hijazi
Faculty of Computing and Informatics, University of Malaysia of Sabah, Kota Kinabalu, Sabah, Malaysia
e-mail: hanafi@ums.edu.my

N. A. Noordin
Faculty of Business, Sharjah Women's Campus, Higher Colleges of Technology, Sharjah, UAE
e-mail: nazima@hct.ac.ae

process of school test analysis, the Malaysian Ministry of Education (MOE) has created a School Exam Analysis System (SAPS) (Kementerian Pendidikan Malaysia, 2016).

There is currently lack of mechanism or approach that can aid MOE in tracking the progress of B40 students holistically. It is now possible to investigate patterns and trends in schoolchildren's academic achievement as well as their lifestyles credits to the development of data mining technology. Using these analyses, top management will examine the growth and academic accomplishment of students, as well as their lifestyle relationships. It also aids in determining what causes B40 community students' performance to improve or depreciate. The study also considers personality qualities, financial stability, social-emotional development at home and school, technology and education, health and well-being, and access to physical and material resources as elements in a student's lifestyle that might affect their academic accomplishment.

Consequently, this paper describes a Data Analytics Pattern Analysis framework in assisting decision-makers to discover and analyze correlation patterns of academic performance in data and associated lifestyles of B40 community pupils effectively. The purpose of this research is to build and adapt correlation techniques for determining any relationship between academic achievement and lifestyle among B40 family schoolchildren. The framework is constructed in four main stages: (1) data collection, (2) data processing, (3) data analytics application, and (4) pattern visualization and evaluation.

The remainder of the paper is organized in the following manner. Section 2 delves into a few related topics, including a review of the literature on a number of B40 community schooling studies and data mining techniques. The modules of the proposed framework for Data Analytics Pattern Analysis are then described in Sect. 3. The experimental setup of the demographic and correlation analysis using a set of survey data, as well as the findings, are presented in Sect. 4. Finally, Sect. 5 concludes with a summary and recommendations for further research.

2 Background

2.1 B40 Community

In general, the household income classifications in Malaysia are B40, M40, and T20. The bottom 40% of Malaysian family income is represented by B40 (income is below RM4,850 per month), the middle 40% is represented by M40 (income between RM4,851 per to RM10,970 per month), and the highest 20% is represented by T20 (income exceeds RM10,971 a month). Recently, the B40 community is expanding and experiencing more reductions in their incomes due to the COVID-19 pandemic. The percentage fall in income for B40 and M40 households was greater than that for T20 households, the income distribution for B40 and M40 households fell to 15.9% in 2020 from 16.0% in 2019 and 36.9% in 2020 from 37.2% in 2019, respectively (Department of Statistics Malaysia, 2020).

According to a few research findings, B40 children face numerous disadvantages. A study carried out by UNICEF (UNICEF, 2018) showed how poverty affects children living in low-cost flats in Kuala Lumpur. Among the study's key findings are that (1) 99.7 percent of children in low-cost flats live in relative and absolute poverty, (2) 15 percent of these children under 5 years old are underweight, (3) 22 percent of the children are undersized, and approximately 23% are either overweight or obese, and (4) while most children aged 7–17 years are enrolled in school, only 50% of children aged 6–12 years went to kindergarten (5) Approximately one-third of all families report that they did not purchase reading materials for their children under the age of 18 years, and four-tenths of all households did not purchase toys for their children under the age of 5 years. Many comparable research efforts and programs are being implemented in Asia to help underprivileged children with their academics in order to improve their economy and, as a result, their lifestyle (Kohyama, 2017; Banerjee, 2016; Khairunnisa & Safri., 2015).

2.2 Education Monitoring System

One of the efforts mentioned in the Eleventh Malaysia Plan for 2016–2020 is to boost the people's economy in order to ensure the country's economy continues to expand. The goal of this project is to enhance economic standards to alleviate poverty and socioeconomic disparities, particularly among low-income households (B40) (Economic Planning Unit, 2016). This research also ranked Malaysia as the 14th best country in terms of educational advancement. This ranking elevates Malaysia's educational system beyond that of the United States, the United Kingdom, and Germany (Kementerian Pendidikan Malaysia, 2017).

The School Examination Analysis System (SAPS) is a system for analyzing school exams in Malaysia. This system was created to save, gather, and analyze data on student exam outcomes in schools. This system was created online and is accessible to all parties, including the Ministry of Education (MOE), the State Education Department, the District Education Office, and the schools. SAPS was created to make it easier for parents to evaluate their child's test results. Parents may follow their child's academic progress via the SAPS revision system, which provides student results slips and scores. SAPS can assess an exam's achievement and identify students' shortcomings because it can create such extensive data analysis. Because each teacher only enters a score into the system and all subject teachers share related exam data too. SAPS, on the other hand, does not provide a deep analysis module that is more effective and secure.

Many accomplishments of various levels of academic performance exist among schools in developing and underdeveloped countries, which are based on differences in socioeconomic background and student lifestyle, resulting in significant differences in student achievement in school (Kohyama, 2017; Ahmad & Sulaiman,

2020). The B40 community's lifestyle and economic challenges have also played a significant influence. Some parents are unconcerned about how their children's schools are run. Furthermore, a person's overall wealth, as well as the obligation of parents to their children in the educational process, has a big influence on carrying out their duties and responsibilities. As a result, the children's academic success is influenced by their lifestyle (Khairunnisa & Safri., 2015; Gopal, 2018; Nohuddin et al., 2021).

2.3 Descriptive Data Mining and Related Works

Data mining (DM) is a data analytics process that uses DM algorithms (Roiger, 2017) or DM tools like RapidMiner, Weka, Orange, SAS, TIBCO, R Studio, and others to find intriguing trends and intelligible patterns (rules) in enormous amounts of data. Predictive and descriptive DM tasks are the two primary categories of DM tasks (Dunham, 2006; Han et al., 2011). Predicting future outcomes (output) based on current data (input) utilizing existing results from other datasets is a common application of a predictive model. For example, a student's performance in an educational institution can be predicted using a variety of academic data factors such as CGPA, GPA, period of study, and so on. Time series analysis, decision trees, and logistic regression are only a few of the common predictive model algorithms identified by researchers in (Joseph, 2022). Weather forecasting, advertising and marketing, healthcare, and education are just a few examples of where predictive models are applied. The descriptive, on the other hand, specifies the properties of the data in a target data set. In other words, descriptive research uses existing data to uncover behavioral patterns or principles and generate new, useful knowledge. For example, using association rule algorithms (ARM), cyberbully behavior patterns can be detected on the Twitter dataset (Zainol et al., 2018), patterns, or rules discovered from the borrowing book record can help the manager recommend books to readers (Yi et al., 2018), student academic performance can be discovered from educational data using ARM (Yuliansyah et al., 2019; Meghji et al., 2018), and many more. Association, clustering, and summarization are some examples of descriptive DM tasks.

Additionally, one of the association descriptive models is correlation analysis. In data mining, correlation analysis is performed to determine the relationship between variables or data attributes using a correlation coefficient measure. Using correlation analysis, a commonly used statistical measure, several studies have successfully revealed fascinating collinear relationships among different features of datasets. This study used correlation analysis to uncover characteristics that have a strong or weak association between lifestyle and academic profiling and analysis.

DM approaches have been used in a variety of studies to evaluate student achievement and performance. Using cluster analysis and regression analysis, Lukkarinen et al. (Lukkarinen et al., 2016) studied the association between student class attendance and learning performance. The experimental dataset was collected from 86 students at a Finnish university in Autumn 2014. Exam points, class attendance, bonus motivation, age, gender, and pre-course are among the six variables in the dataset. They separated the data into three groups. Exam scores had a positive correlation with three other variables, such as class attendance, bonus motivation, and gender, according to the descriptive analysis of all variables in cluster 2 (female). According to a similar study (Fadelelmoula, 2018), there is a favorable relationship between student class attendance and final test performance in all courses.

Sani et al. (2020) conducted a comparison study utilizing machine learning technique. Three different classifiers were employed in their study to predict attrition of B40 students enrolled in undergraduate programs: Decision Tree (DT), Random Forest (RF), and Artificial Neural Network (ANN). The experiment dataset was provided by the MOE. There was a total of 44,406 records, each having a unique quality. Two new attributes (age and class) were added to the dataset. The descriptive findings indicated that 23 characteristics had a significant link with the class label when the Chi-Square test was used. On the other hand, the attributes place of birth and family income were omitted from the dataset based on the findings of the Cramer's V test because they were less significant to class label. Finally, the dataset selected has 28,844 records including ten properties (including class label). Similarly, Random Forest was chosen as the best model.

Abdol et al. (2021) investigated the correlations between student engagement on an e-learning platform and classroom performance. The data was acquired via Kaggle and analyzed using the K-means algorithm. The associations between three key variables were examined in this study: student conduct (engagement), peer reaction, and skill grades. Correlations between these variables were visualized using scatter plots, heatmaps, bar graphs, and other techniques.

In another study, Das et al. (2020) compared four (4) classification techniques for detecting student academic performance: decision tree, Naive Bayes, K-Nearest Neighbor (KNN), and Artificial Neural Network (ANN). The dataset was downloaded from the Kaggle website, and three (3) feature selection algorithms were utilized in this study: chi-Square, Euclidean distance, and correlation. Variables like libraryMatrials, resultAssessment, classroomFacilities, SGPA, internetSpeed, and indoorandoutdoorMedicalFacalities were found to be more correlated with student academic success. The results revealed that ANN performed better when the number of classes was reduced.

2.4   Factors That Affect Academic Performance

This study focuses on seven factors of the student's lifestyle that can affect their academic achievement which are personality traits, financial stability, social-emotional development at home and school, technology and education, health and well-being, and access to physical and material resources.

2.4.1 Personality Traits

Personality traits represent a person's self-behavior or self-personality, which includes hobbies and daily activities. It also encompasses a person's actions, feelings, and decisions. A study focused on determining whether emotional knowledge had a beneficial or detrimental impact on a student's behavior and academic performance. The findings suggested that a lack of emotional awareness might lead to negative qualities in children as well as poor academic achievement. It can also have an impact on the children's mood, focus, and motivation to do well in school (Steinmayr et al., 2019).

2.4.2 Financial Stability

Another aspect is the financial stability of the parents. It is one of the factors that can have a direct impact on a child's academic achievement. Parents with more financial stability can provide for their children's needs, particularly in terms of schooling. According to a study, the relationship between family resources and children's educational attainment has an impact. Students who grew up in a low-income household where their parents struggled financially can experience early-life stress, which affects their brain development and functioning (Hair et al., 2015; Blair & Raver, 2016).

2.4.3 Social-Emotional Development (at Home)

The importance of family affection in the development of children cannot be overstated. They are more prone to notice and react to any changes in the family, whether favorable or negative. A study on the relationship between parental participation and student academic achievement looked at how well parents monitored their children's homework, how supportive they were of their children, and how often they participated in school events or activities. Parental participation is critical for children to succeed academically and obtain better results (Ntekane, 2018).

2.4.4 Social Emotional Development (at School)

A review investigated into the nature of stress among university and secondary school students, as well as the sources of stress in their learning process and academic performance. The study gathered all of the student's demographic information, personal information, and stress-related questions. It was discovered that academic issues such as exams and trouble understanding the contents of learning subjects cause students to be extremely stressed, which can negatively impact their academic performance (Othman et al., 2013; Vestad & Tharaldsen, 2021).

2.4.5   Technology and Education

Another key aspect that can affect a student's academic success is technology and education. Students are now exposed to ICT education in order for them to achieve well in their schoolwork and exams. The impact of ICT education on pupils' academic performance was examined in a study. The study looked at accessing computers off campus, accessing the Internet off campus, and other topics, with the conclusion that utilizing computers at home helps students achieve better academic results because they can readily access the Internet to complete their work (Ansari & Khan, 2020).

2.4.6   Health and Well-Being

A study (Maniaci et al., 2021) found that students' academic achievement can be affected by health and behaviors. The study shows that students that achieve a higher BMI faced a lower self-esteem, depression, and a poor grade. Students with high BMI are being analyzed that they eat less nutrition food and lack in doing physical activities. This has been given an impact on the students which lead to low academic achievement among them.

2.4.7   Access to Physical and Material Resources

A study (Muhmmad et al., 2021) focused on examining the link between academic achievement and a student's home environment, with elements such as home learning facilities, parental encouragement, and others. The findings suggest that providing a child with the necessary learning resources and parental encouragement can help them attain academic success.

3   Methodology

This section explains about the methodology used to collect data and experimental setup of the analysis framework.

3.1   B40 Schoolchildren Survey Data

The information was gathered over the course of 2 days by distributing question-naires to Sekolah Kebangsaan Sri Jerlok, Kajang. About 50 students completed

II) FAKTOR DI RUMAH (Social-emotional development - Home):

Soalan:
Anda sering mengalami masalah tekanan/stress di rumah.
Anda perlu membantu keluarga selepas sekolah. Contohnya, membantu ibubapa untuk berniaga atau bekerja.
Anda sering melihat pergaduhan ibubapa atau adik-beradik ketika di dalam rumah.
Anda tidak mempunyai mentor (ibubapa)/guru/tutor untuk mengajar anda ketika di rumah.
Anda sering pergi beriadah dan melakukan pelbagai aktiviti bersama keluarga seperti berkelah atau bermain badminton.
Ibu bapa anda mengalami masalah kesihatan yang serius seperti sakit jantung.

III) FAKTOR DI SEKOLAH (Social-emotional Development- School):

Soalan:
Anda mempunyai masalah kenderaan untuk menghantar dan pulang ke sekolah.
Anda mengalami masalah tekanan/stress di sekolah.
Anda mengalami kesukaran untuk belajar kerana tidak tahu teknik dan cara belajar yang betul.
Anda sering memberikan tumpuan yang penuh ketika waktu pembelajaran.
Anda mempunyai masalah dalam menyiapkan kerja khusus sekolah kerana tidak mempunyai PC/laptop.
Anda sering menghadiri ke kelas setiap hari.

Fig. 1 Part of survey questionnaires which include the seven factors of their lifestyles

questionnaires, aged between 9 and 10 years. These pupils come from B40 community families.

The surveys for students are separated into two sections: Demographic and Psychographic questionnaires. Name, Age, Gender, Family Status, Class Level, Number of Siblings, and Race are all included in Demographic Data. Time with parents, sleeping time, revision time, attending tuition class, Mother's employment, Father's occupation, Physical activity, and seven variables of student's lives that might impact academic achievement are all included in Psychographic Data. Figure 1 illustrates the set of questions for social-emotional development factors in the survey. The questions cover social problems or emotional issues faced by the B40 schoolchildren at home and at school.

The aim of the research is to focus on the association between the students' lifestyle and their academic achievement in order to complete the study. Personal traits, financial stability, social-emotional development at home and school, technology and education, health and well-being, and access to physical and material resources are all included in the surveys. The survey is conducted in the Malay language.

3.2   Data Analytics Pattern Analysis Framework

In this study, there are several integrated modules involved in generating the lifestyle and academic profiling and analysis. Figure 2 depicts the Data Analytics Pattern Analysis Framework (DAPAF) which consists of four (4) modules which are (1) B40 schoolchildren survey data, (2) data preprocessing, (3) correlation pattern analyzer, and (4) visualizations.

Initially, as mentioned earlier, data was collected through a survey session done at Sekolah Kebangsaan Sri Jerlok, Kajang. Around 50 students, ages ranging from 9 to 10 years, answered 44 questionnaires which were divided into 2 elements, demographic and psychographic questionnaires. Each survey question is rated on a five-point Likert scale, with 1 indicating disagree and 5 indicating strongly agree. From the questionnaires, the dimensions of data collection were 50 records x 44 attributes (answers).

Then, these survey forms undertake a data preprocessing module which comprises of data preparation and data cleaning. Data is tabulated according to attribute

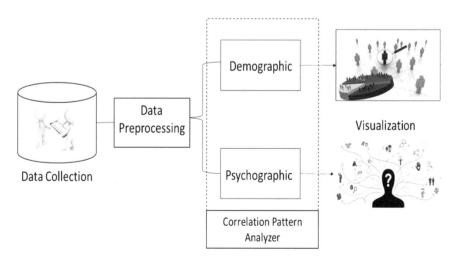

Fig. 2 Data analytics pattern analysis framework

Fig. 3 Correlation analysis process

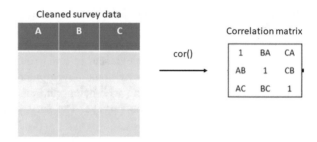

grouping, demographic and psychographic. The answers are transformed into 1 and 0 values. Data is also filtered for missing values and replaced with global constant value.

The next module analyses and extracts interesting patterns from the cleaned survey data. The module is developed based on demographic analysis and correlation techniques using R. R is a programming language that we use to explore, analyze, and comprehend the collected data. The algorithms are applied for identifying and describing the relationship between demographic and psychographic attributes from the survey. We evaluate data directly from survey data sets. The data attributes are divided into personal and time management details and survey answers for personal traits, financial ability for extra classes, social-emotional development activities at home and school, technology privileges, health and well-being, conducive learning environment at home, and academic performance. The demographic analysis includes the frequency counts of each data attribute, and they are differentiated based on gender.

In the correlation pattern analyzer, details of data ranges (survey answers) are also included to illustrate the trend with each dataset profile. This module also extracts correlation patterns based on personal, academic, financial, and social traits from the survey data attributes. This is to measure the closeness of association of the attribute values. Correlation matrix analysis is a powerful tool for determining the relationships between variables. This study uses cor(), a R function for quickly calculating and viewing a correlation matrix. The representation of correlation coefficient and p-values of the correlations are displayed (Fig. 3). As a result, the variables are reordered according to the degree of correlation, making it easier to spot the most closely related data attributes in the correlation matrices.

Lastly, in the visualization module, the generated patterns are then presented and visualized in histogram and correlation matrices.

4   Results and Discussion

In this section, we discuss on data analysis, demographic, and correlation of the survey data so as to investigate the relationship between lifestyle and academic performance of B40 schoolchildren.

4.1 Demographic Analysis

Demographic analysis is useful in understanding characteristics of B40 schoolchildren on gender, family size, parents' marital status, and time spent on family, study, and leisure activities.

Figure 4 depicts the gender distribution of B40 schoolchildren who answered the survey. About 56% are male and 44% are female. Then Fig. 5 describes the size of sibling members. Majority of the schoolchildren has 4 siblings (34%), followed by 2 siblings (20%), 3 siblings (18%), 5 siblings (16%), 6 siblings (10%), and 1 sibling (2%).

Figure 6 illustrates the marital status of their parents. Ninety percent of the parents are married and 10% of the parents are either divorced or widowed. Whereas Fig. 7

Fig. 4 Schoolchildren gender

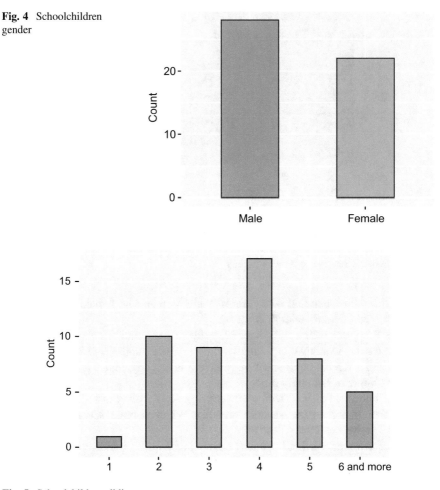

Fig. 5 Schoolchildren siblings

Fig. 6 Schoolchildren
parent's marital status

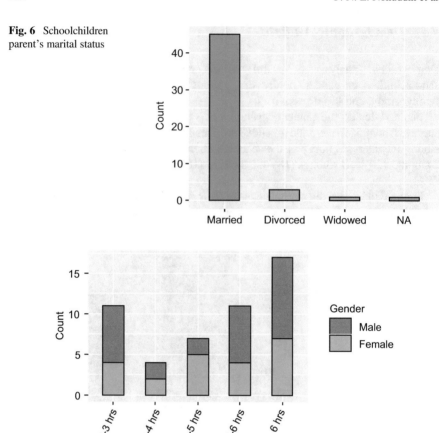

Fig. 7 Schoolchildren time spent with parents

explains their time spending with parents. Majority male and female schoolchildren spent more than 6 hours with their parents.

Figure 8 depicts their study time. Majority of male and female schoolchildren spent 1–2 h a day to study their schoolwork. Two male pupils spent 3–4 h and 4–5 h to study, and 1 female pupil spent more than 5 h to study, and 1 female pupil was unsure the hours of her study time.

Figure 9 portrays their activities between 5 pm and 7 pm, most of them preferred to have free activities (no specific pursuit). Whereas male schoolchildren also preferred to have sport activities.

Malaysian primary school syllabus consists of 6 core subjects which are Mathematics, English, Malay Language, History, Science, and Islamic Studies. From Fig. 10, about 13 pupils acquired academic results between 6A and 4A in the examination. Majority of the respondents (22 pupils) gained results of 3A-2A in

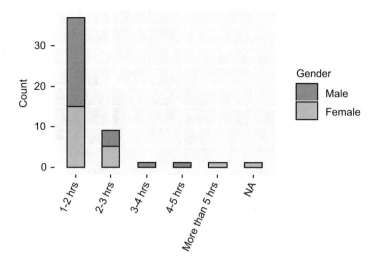

Fig. 8 Schoolchildren study time

Fig. 9 Schoolchildren activities between 5 pm and 7 pm

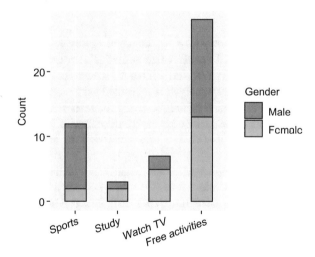

their examinations. However, the rest of them got 1A and below academic results. However, it is interesting to find that 90% of the B40 schoolchildren did not attend any extra tuition classes other than classes at school.

4.2 Relationship Patterns Using Correlation Analysis

The pattern examination goes deeper with correlation analysis in order to identify relationships and correlations between the data attributes which are survey answers

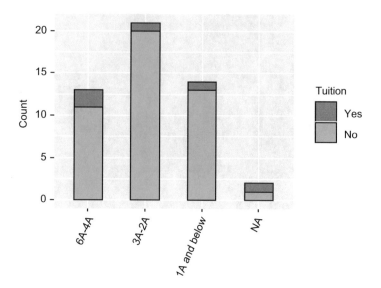

Fig. 10 Academic results

on personal traits (C1-C5), financial ability for attending extra classes (D1-D5), social-emotional development activities at home (E1-E6), health, food intake, and well-being activities (H1-H7) and academic result (K2). Using the correlation technique, pattern analysis is presented using correlation matrices. Diagonal element K refers to number of unique levels for each item. The numerical value in each row represents the association measure from item X to item Y. The value in each column represents the association measure (week or strong association) from item Y to item X.

Figure 11 depicts correlation matrix in which C1 to C5 represent personality traits and K2 is academic results. On average, weak association between items in personality traits. A good sign such that each item can be said to be unique from each other. The values show that weak association between personality traits and academic results which can be deduced that regardless of a positive or negative schoolchildren's personal character, they still achieved good academic performance.

In Fig. 12, items D1 to D5 represent financial stability of B40 families versus K2 which is academic results. From the matrix, a strong association, value = 0.23, is shown from D3 (Supplement for mental development) to D2 (Expenses for activity books). This could indicate their parents maybe spend money on children mental and knowledge development even though D1 revealed that majority of the respondents did not attend extra classes due to financial stability and ability (match with Fig. 10 too). Nevertheless, there is a weak association between financial stability and academic results. No association between D4 with the rest. D4 did not contribute to the outcome of K2.

Figure 13 illustrates items E1 to E6 for Social-emotional development activities (home) versus K2 for academic results. Also, a fairly strong association,

Fig. 11 Personal traits versus academic

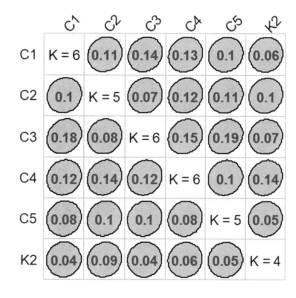

Fig. 12 Financial stability versus academic

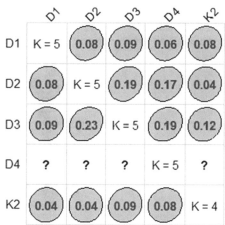

value = 0.22, from E4 (Parents or family members act as a mentor for schoolwork) to E3 (Witness less arguments between parents) can be seen in the matrix. This could indicate harmonized home atmosphere. Overall, there is a weak association between social-emotional (home) and academic results.

Figure 14 depicts correlation matrix in which H1 to H7 represents health, food intake, and well-being activities and K2 is academic results. On average, weak association between items in health, food awareness, and well-being activities. However, a fairly strong association, value = 0.24, from H7 (Visit doctor when they are not well) to H6 (Daily intake of vitamin c) can be seen in the matrix. This could indicate B40 families have a good health, food intake, and well-being

Fig. 13 Social-emotional
versus academic

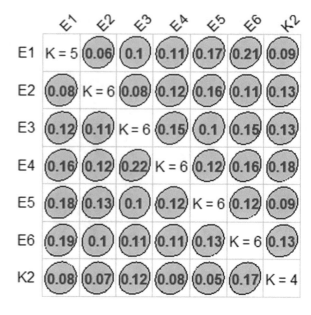

Fig. 14 Health and well-
being versus academic

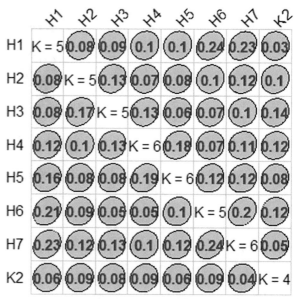

awareness. Overall, there is a weak association between health, food intake and well-being activities, and academic results.

In general, pattern analysis utilizing the correlation technique reveals both internal and external association values between the survey's selected qualities. Correlation analysis can be used to discover significant and meaningful correlations between various personal trait items (C1-C5), financial ability to attend additional classes

(D1-D5), social-emotional development activities at home (E1-E6), health, food intake, and well-being activities (H1-H7), and academic result (K2). Despite the B40 community's constant claims and concerns about financial and educational sustainability, B40 schoolchildren do admirably in educational programs at their schools. Information about these relationships can elucidate interdependencies and provide new insights. As a result, it benefits stakeholders such as the government and educational institutions. Likewise, Malaysia aspires to become a developed nation by 2020, and in order to do so, the country's previous methods to poverty and inequality reduction must be maintained. To ensure that no one is left behind in the pursuit of the Sustainable Prosperity Goals, steps have been planned or executed to ensure Malaysians have access to the country's wealth and development.

5 Conclusion

Finally, the Data Analytics Pattern Analysis Framework was developed to assess survey data in order to establish the link between B40 community school children's lifestyle and academic success. The framework is constructed in phases: (1) survey data collecting, (2) data processing and preparation, (3) correlation algorithm application, and (4) visualization of patterns. Top management can assess students' growth and academic accomplishments, as well as their lifestyle relationships, using these analyses. It also enables the investigator to figure out what causes B40 community students' grades to improve or drop. Personality traits, socioeconomic stability, social-emotional development at home and school, technology, and education, health and well-being, and access to physical and material resources are all factors that could affect a student's academic success, according to the study.

Acknowledgments This research is fully supported by GUP grant, GUP-2018-076. The authors fully acknowledged the Ministry of Higher Education (MOHE) and Universiti Kebangsaan Malaysia for the research fund which makes this study viable and effective.

References

Abdo, A. M., Rasid, N. M. A., Badli, N. A. H. M., Sulaiman, S. N. A., Wani, S., & Zainol, Z. (2021). Student's performance based on E-learning platform behaviour using clustering techniques. *International Journal on Perceptive and Cognitive Computing, 7*(1), 72–78.

Ahmad, N., & Sulaiman, N. (2020). Demographic and socio-economic characteristics, household food security status and academic performance among primary school children in north Kinta, Perak, Malaysia. *Malaysian Journal of Medicine and Health Sciences, 16*(suppl. 6), 26–33.

Ansari, J. A. N., & Khan, N. A. (2020). Exploring the role of social media in collaborative learning the new domain of learning. *Smart Learn. Environ., 7*, 9. https://doi.org/10.1186/s40561-020-00118-7

Banerjee, P. (2016). A systematic review of factors linked to poor academic performance of disadvantaged students in science and maths in schools. *Cogent Education, 3*, 1178441.

Blair, C., & Raver, C. C. (2016). Poverty, stress, and brain development: New directions for prevention and intervention. *Academic Pediatrics, 16*(3 Suppl), S30–S36. https://doi.org/10.1016/j.acap.2016.01.010

Das, D., et al. (2020). A comparative analysis of four classification algorithms for university students performance detection. In *ECCE 2019* (pp. 415–424). Springer.

Department of Statistics Malaysia. (2020). Household income estimates and incidence of poverty report, Malaysia. DOSM. Retrieved February 20, 2022, from https://www.dosm.gov.my/v1/index.php?r=column/cthemeByCat&cat=493&bul_id=VTNHRkdiZkFzenBNd1Y1dmg2UUlrZz09&menu_id=amVoWU54UTl0a21NWmdhMjFMMWcyZz09

Dunham, M. H. (2006). *Data mining: Introductory and advanced topics*. Pearson Education India.

Economic Planning Unit. (2016). P. M. D. Elevating B40 Households towards a Society.

Fadelelmoula, T. (2018). The impact of class attendance on student performance. *International Research Journal of Medicine and Medical Sciences, 6*(2), 47–49.

Gopal, P. S. (2018). Poverty measurement revisited from a multidimensional perspective among Universiti Sains Malaysia's B40 poor students. *Geografia-Malaysian Journal of Society and Space, 14*(4), 299–307.

Hair, N. L., Hanson, J. L., Wolfe, B. L., & Pollak, S. D. (2015). Association of child poverty, brain development, and academic achievement. *JAMA Pediatrics, 169*(9), 822–829.

Han, J., Pei, J., & Kamber, M. (2011). *Data mining: Concepts and techniques*. Elsevier.

Joseph, M. C. (2022). Predictive modeling. TechTarget. Retrieved February 20, from https://searchenterpriseai.techtarget.com/definition/predictive-modeling

Kementerian Pelajaran Malaysia. (2016). Sistem Analisis Peperiksaan Sekolah. https://sapsnkra.moe.gov.my/ibubapa2/index.php.

Kementerian Pendidikan Malaysia. (2017). Pencapaian Malaysia Dalam Indeks Persaingan Global Dalam Bidang Pendidikan 2010–2013. Retrieved February 20, 2022, from https://www.moe.gov.my/index.php/my/kenyataan-media-akhbar-2012/540-pencapaian-malaysia-dalam-indeks-persaingan-global-dalam-bidang-pendidikan?templateStyle=9

Khairunnisa, F. S., & Safri. (2015). Hubungan Gaya Hidup Dengan Prestasi Akademik Mahasiswa Keperawatan Universitas Riau. *Jurnal Online Mahasiswa, 2*(2), 2015.

Kohyama, J. (2017). Self-reported academic performance and lifestyle habits of school children in Japan. *International Journal of Child Health and Nutrition, 6*(3), 90–97.

Lukkarinen, A., Koivukangas, P., & Seppälä, T. (2016). Relationship between class attendance and student performance. *Procedia-Social and Behavioral Sciences, 228*, 341–347.

Maniaci, G., La Cascia, C., Giammanco, A., et al. (2021). The impact of healthy lifestyles on academic achievement among Italian adolescents. *Current Psychology*. https://doi.org/10.1007/s12144-021-01614-w

Meghji, A. F., Mahoto, N. A., Unar, M. A., & Shaikh, M. A. (2018). Analysis of student performance using EDM methods. In *2018 5th International Multi-Topic ICT Conference (IMTIC)* (pp. 1–7). IEEE.

Muhmmad, Y., Liu, C., Khalid, S., & Bakar, A. (2021). Effect of home environment on students'. *Academic Achievements at Higher Level, 20*(358–369). https://doi.org/10.17051/ilkonline

Nohuddin, P. N., Zainol, Z., & Hijazi, M. H. A. (2021). Study of B40 Schoolchildren Lifestyles and Academic Performance using Association Rule Mining. *Annals of Emerging Technologies in Computing (AETiC), 5*(5), 60–68.

Ntekane, A. (2018). *Parental involvement in education* (Vol. 10.13140/RG.2.2.36330.21440).

Othman, C. N., Farooqui, M., Yusoff, M. S. B., & Adawiyah, R. (2013). Nature of stress among health science students in a Malaysian university. *Procedia - Social and Behavioral Sciences., 105*, 249–257. https://doi.org/10.1016/j.sbspro.2013.11.026

Roiger, R. J. (2017). *Data mining: A tutorial-based primer*. CRC Press.

Sani, N. S., Nafuri, A. F. M., Othman, Z. A., Nazri, M. Z. A., & Mohamad, K. N. (2020). Drop-out prediction in higher education among B40 students. *International Journal of Advanced Computer Science and Applications, 11*(11), 550–559.

Steinmayr, R., Weidinger, A. F., Schwinger, M., & Spinath, B. (2019). The importance of students' motivation for their academic achievement – Replicating and extending previous findings. *Frontiers in Psychology, 10,* 1730.

UNICEF. (2018). Children without: A study of urban child poverty in low cost flats in Kuala Lumpur. UNICEF Malaysia & DM Analytics. Retrieved June 1, 2021, from https://www.unicef.org/malaysia/reports/children-without

Vestad, L., & Tharaldsen, K. B. (2021). Building social and emotional competencies for coping with academic stress among students in lower secondary school. *Scandinavian Journal of Educational Research, 0*(0), 1–15.

Yi, K., Chen, T., & Cong, G. (2018). *Library personalized recommendation service method based on improved association rules.* Library Hi Tech.

Yuliansyah, H., Hafsah, I. A., & Umar, R. (2019). Discovering Meaningful Pattern of Undergraduate Students Data using Association Rules Mining. In *2019 Ahmad Dahlan International Conference Series on Engineering and Science (ADICS-ES 2019)* (pp. 13–17). Atlantis Press.

Zainol, Z., Wani, S., Nohuddin, P. N. E., Noormanshah, W. M. U., & Marzukhi, S. (2018). Association analysis of cyberbullying on social media using Apriori algorithm. *International Journal of Engineering & Technology, 7*(4.29), 72–75. https://doi.org/10.14419/ijet.v7i4.29.21847

Printed in the
by Baker & Ta

Printed in the United States
by Baker & Taylor Publisher Services